advancing learning, changing lives

BTEC National
Engineering

Study Guide

A PEARSON COMPANY

BTEC National Study Guide: Engineering

Published by:
Edexcel Limited
One90 High Holborn
London WC1V 7BH
www.edexcel.org.uk

Distributed by:
Pearson Education Limited
Edinburgh Gate
Harlow
Essex CM20 2JE

First published 2007

ISBN 978-1-84690-213-0

Project managed and typeset by Hart McLeod, Cambridge
Printed in Great Britain by Henry Ling Ltd., at the Dorset Press, Dorchester, Dorset

Cover image ©Richard Osbourne/Blue Pearl Photography/Alamy

The publisher's policy is to use paper manufactured from sustainable forests.

All reasonable efforts have been made to trace and contact original copyright owners.

This material offers high quality support for the delivery of Edexcel qualifications.
This does not mean that it is essential to achieve any Edexcel qualification, nor does it mean that this is the only suitable material available to support any Edexcel qualification. No Edexcel-published material will be used verbatim in setting any Edexcel assessment and any resource lists produced by Edexcel shall include this and other appropriate texts.

Contents

PREFACE

If you've already followed a BTEC First programme, you will know that this is an exciting way to study; if you are fresh from GCSEs you will find that from now on you will be in charge of your own learning. This guide has been written specially for you, to help get you started and then succeed on your BTEC National course.

The **Introduction** concentrates on making sure you have all the right facts about your course at your fingertips. Also, it guides you through the important skills you need to develop if you want to do well including:

- managing your time
- researching information
- preparing a presentation.

Keep this by your side throughout your course and dip into it whenever you need to.

The **Activities** give you tasks to do on your own, in a small group or as a class. They will help you internalise your learning and then prepare for assessment by practising your skills and showing you how much you know. These activities are not for assessment.

The sample **Marked Assignments** show you what other students have done to gain Pass, Merit or Distinction. By seeing what past students have done, you should be able to improve your own grade.

Your BTEC National will cover six, twelve or eighteen units depending on whether you are doing an Award, Certificate or Diploma. In this guide the activities cover sections from Unit 2 – Communication for Technicians, Unit 4 – Mathematics for Technicians, Unit 5 – Electrical and Electronic Principles and Unit 6 – Mechanical Principles and Applications. These units underpin your study of Engineering.

Because the guide covers only four units, it is essential that you do all the other work your tutors set you. You will have to research information in textbooks, in the library and on the Internet. You should have the opportunity to visit local organisations and welcome visiting speakers to your institution. This is a great way to find out more about your chosen vocational area – the type of jobs that are available and what the work is really like.

This Guide is a taster, an introduction to your BTEC National. Use it as such and make the most of the rich learning environment that your tutors will provide for you. Your BTEC National will give you an excellent base for further study, a broad understanding of engineering and the knowledge you need to succeed in the world of work. Remember, thousands of students have achieved a BTEC National and are now studying for a degree or at work, building a successful career.

INTRODUCTION

SEVEN STEPS TO SUCCESS ON YOUR BTEC NATIONAL

You have received this guide because you have decided to do a BTEC National qualification. You may even have started your course. At this stage you should feel good about your decision. BTEC Nationals have many benefits – they are well-known and respected qualifications, they provide excellent preparation for future work or help you to get into university if that is your aim. If you are already at work then gaining a BTEC National will increase your value to your employer and help to prepare you for promotion.

Despite all these benefits though, you may be rather apprehensive about your ability to cope. Or you may be wildly enthusiastic about the whole course! More probably, you are somewhere between the two – perhaps quietly confident most of the time but sometimes worried that you may get out of your depth as the course progresses. You may be certain you made the right choice or still have days when your decision worries you. You may understand exactly what the course entails and what you have to do – or still feel rather bewildered, given all the new stuff you have to get your head around.

Your tutors will use the induction sessions at the start of your course to explain the important information they want you to know. At the time, though, it can be difficult to remember everything. This is especially true if you have just left school and are now studying in a new environment, among a group of people you have only just met. It is often only later that you think of useful questions to ask. Sometimes, misunderstandings or difficulties may only surface weeks or months into a course – and may continue for some time unless they are quickly resolved.

This student guide has been written to help to minimise these difficulties, so that you get the most out of your BTEC National course from day one. You can read through it at your own pace. You can look back at it whenever you have a problem or query.

This Introduction concentrates on making sure you have all the right facts about your course at your fingertips. This includes a **Glossary** (on page 32) which explains the specialist terms you may hear or read – including words and phrases highlighted in bold type in this Introduction.

The Introduction also guides you through the important skills you need to develop if you want to do well – such as managing your time, researching information and preparing a presentation; as well as reminding you about the key skills you will need to do justice to your work, such as good written and verbal communications.

Make sure you have all the right facts

5

- Use the PlusPoint boxes in each section to help you to stay focused on the essentials.

- Use the Action Point boxes to check out things you need to know or do right now.

- Refer to the Glossary (on page 32) if you need to check the meaning of any of the specialist terms you may hear or read.

Remember, thousands of students have achieved BTEC National Diplomas and are now studying for a degree or at work, building a successful career. Many were nervous and unsure of themselves at the outset – and very few experienced absolutely no setbacks during the course. What they did have, though, was a belief in their own ability to do well if they concentrated on getting things right one step at a time. This Introduction enables you to do exactly the same!

STEP ONE

UNDERSTAND YOUR COURSE AND HOW IT WORKS

What is a BTEC qualification and what does it involve? What will you be expected to do on the course? What can you do afterwards? How does this National differ from 'A' levels or a BTEC First qualification?

All these are common questions – but not all prospective students ask them! Did you? And, if so, did you really listen to the answers? And can you remember them now?

If you have already completed a BTEC First course then you may know some of the answers – although you may not appreciate some of the differences between that course and your new one.

Let's start by checking out the basics.

- All BTEC National qualifications are **vocational** or **work-related**. This doesn't mean that they give you all the skills that you need to do a job. It does mean that you gain the specific knowledge and understanding relevant to your chosen subject or area of work. This means that when you start in a job you will learn how to do the work more quickly and should progress further. If you are already employed, it means you become more valuable to your employer. You can choose to study a BTEC National in a wide range of vocational areas, such as Business, Health and Social Care, IT, Performing Arts and many others.

- There are three types of BTEC National qualification and each has a different number of units.

  - The BTEC National Award usually has 6 units and takes 360 **guided learning hours (GLH)** to complete. It is often offered as a part-time or short course but you may be one of the many students doing an Award alongside A-levels as a full-time course. An Award is equivalent to one 'A' level.

  - The BTEC National Certificate usually has 12 units and takes 720 GLH to complete. You may be able to study for the Certificate on a part-time or full-time course. It is equivalent to two 'A' levels.

– The BTEC National Diploma usually has 18 units and takes 1080 GLH to complete. It is normally offered as a two-year full-time course. It is equivalent to three 'A' levels.

These qualifications are often described as **nested**. This means that they fit inside each other (rather like Russian dolls!) because the same units are common to them all. This means that if you want to progress from one to another you can do so easily by simply completing more units.

■ Every BTEC National qualification has a set number of **core units**. These are the compulsory units every student must complete. The number of core units you will do on your course depends upon the vocational area you are studying.

■ All BTEC National qualifications also have a range of **specialist units** from which you may be able to make a choice. These enable you to study particular areas in more depth.

■ Some BTEC National qualifications have **specialist core units**. These are mandatory units you will have to complete if you want to follow a particular pathway in certain vocational areas. Engineering is an example of a qualification with the over-arching title, Engineering, which has a set of core units that all students must complete. Then, depending what type of engineering a student wants to follow, there are more specialist core units that must be studied.

■ On all BTEC courses you are expected to be in charge of your own learning. If you have completed a BTEC First, you will already have been introduced to this idea, but you can expect the situation to be rather different now that you are working at BTEC National level. Students on a BTEC First course will be expected to need more guidance whilst they develop their skills and find their feet. In some cases, this might last quite some time. On a BTEC National course you will be expected to take more responsibility for yourself and your own learning almost from the outset. You will quickly be expected to start thinking for yourself. This means planning what to do and carrying out a task without needing constant reminders. This doesn't mean that your tutor won't give you help and guidance when you need it. It does mean, though, that you need to be 'self-starting' and to be able to use your own initiative. You also need to be able to assess your own performance and make improvements when necessary. If you enjoy having the freedom to make your own decisions and work at your own pace then you will welcome this type of learning with open arms. However, there are dangers! If you are a procrastinator (look up this word if you don't know what it means!) then it's quite likely you will quickly get in a muddle. In this case read Step 3 – Use your time wisely – very carefully indeed!

■ The way you are assessed and graded on a BTEC course is different from an 'A' level course, although you will still obtain UCAS points which you need if you want to go to university. You can read about this in the next section.

PLUSPOINTS

+ You can usually choose to study part-time or full-time for your BTEC National and do an Award, Certificate or Diploma and progress easily from one to the other.

+ You will study both core units and specialist units on your course.

+ When you have completed your BTEC course you can get a job (or **apprenticeship**), use your qualification to develop your career and/or continue your studies to degree level.

+ You are responsible for your own learning on a BTEC course. This prepares you for life at work or at university when you will be expected to be self-starting and to use your own initiative.

ACTION POINTS

✓ Check you know whether you are studying for an Award, Certificate or Diploma and find out the number of units you will be studying for your BTEC National qualification.

✓ Find out which are core and which are specialist units, and which specialist units are offered at your school or college.

✓ Check out the length of your course and when you will be studying each unit.

✓ Explore the Edexcel website at www.edexcel.org.uk. Your first task is to find what's available for your particular BTEC National qualification. Start by finding National qualifications, then look for your vocational area and check you are looking at the 2007 schemes. Then find the specification for your course. Don't print this out – it is far too long. You could, of course, save it if you want to refer to it regularly or you could just look through it for interest and then bookmark the pages relating to your qualification for future reference.

✓ Score yourself out of 5 (where 0 is awful and 5 is excellent) on each of the following to see how much improvement is needed for you to become responsible for your own learning!

Being punctual; organisational ability; tidiness; working accurately; finding and correcting own mistakes; solving problems; accepting responsibility; working with details; planning how to do a job; using own initiative; thinking up new ideas; meeting deadlines.

✓ Draw up your own action plan to improve any areas where you are weak. Talk this through at your next individual **tutorial**.

STEP TWO

UNDERSTAND HOW YOU ARE ASSESSED AND GRADED – AND USE THIS KNOWLEDGE TO YOUR ADVANTAGE!

If you already have a BTEC First qualification, you may think that you don't need to read this section because you assume that BTEC National is simply more of the same. Whilst there are some broad similarities, you will now be working at an entirely different level and the grades you get for your work could be absolutely crucial to your future plans.

Equally, if you have opted for BTEC National rather than 'A' level because you thought you would have less work (or writing) to do then you need to read this section very carefully. Indeed, if you chose your BTEC National because you thought it would guarantee you an easy life, you are likely to get quite a shock when reality hits home!

It is true that, unlike 'A' levels, there are no exams on a BTEC course. However, to do well you need to understand the importance of your assignments, how these are graded and how these convert into unit points and UCAS points. This is the focus of this section.

Your assignments

On a BTEC National course your learning is assessed by means of **assignments** set by your tutors and given to you to complete throughout your course.

■ Your tutors will use a variety of **assessment methods**, such as case studies, projects, presentations and shows to obtain evidence of your skills and knowledge to date. You may also be given work-based or **time-constrained** assignments – where your performance might be observed and assessed. It will depend very much on the vocational area you are studying (see also page 16).

■ Important skills you will need to learn are how to research information (see page 25) and how to use your time effectively, particularly if you have to cope with several assignments at the same time (see page 12). You may also be expected to work cooperatively as a member of a team to complete some parts of your assignments – especially if you are doing a subject like Performing Arts – or to prepare a presentation (see page 26).

■ All your assignments are based on **learning outcomes** set by Edexcel. These are listed for each unit in your course specification. You have to meet *all* the learning outcomes to pass the unit.

Your grades

On a BTEC National course, assignments that meet the learning outcomes are graded as Pass, Merit or Distinction.

■ The difference between these grades has very little to do with how much you write! Edexcel sets out the **grading criteria** for the different grades in a **grading grid**. This identifies the **higher-level skills** you have to demonstrate to earn a higher grade. You can find out more about this, and read examples of good (and not so good) answers to assignments at Pass, Merit and Distinction level in the marked assignments section starting on page 119. You will also find out more about getting the best grade you can in Step 5 – Understand your assessment – on page 16.

■ Your grades for all your assignments earn you **unit points**. The number of points you get for each unit is added together and your total score determines your final grade(s) for the qualification – again either Pass, Merit or Distinction. You get one final grade if you are taking a BTEC National Award, two if you are taking a BTEC National Certificate and three if you are taking a BTEC National Diploma.

■ Your points and overall grade(s) also convert to **UCAS points** which you will need if you want to apply to study on a degree course. As an example, if you are studying a BTEC National Diploma, and achieve three final pass grades you will achieve 120 UCAS points. If you achieve three final distinction grades the number of UCAS points you have earned goes up to 360.

■ It is important to note that you start earning both unit and UCAS points from the very first assignment you complete! This means that if you take a long time to settle into your course, or to start working productively, you could easily lose valuable points for quite some time. If you have your heart set on a particular university or degree course then this could limit your choices. Whichever way you look at it, it is silly to squander potentially good grades for an assignment and their equivalent points, just because you didn't really understand what you had to do – which is why this guide has been written to help you!

■ If you take a little time to understand how **grade boundaries** work, you can see where you need to concentrate your efforts to get the best final grade possible. Let's give a simple example. Chris and Shaheeda both want to go to university and have worked hard on their BTEC National Diploma course. Chris ends with a total score of 226 unit points which converts to 280 UCAS points. Shaheeda ends with a total score of 228 unit points – just two points more – which converts to 320 UCAS points! This is because a score of between 204 and 227 unit points gives 280 UCAS points, whereas a score of 228 – 251 points gives 320 UCAS points. Shaheeda is pleased because this increases her chances of getting a place on the degree course she wants. Chris is annoyed. He says if he had known then he would have put more effort into his last assignment to get two points more.

It is always tempting to spend time on work you like doing, rather than work you don't – but this can be a mistake if you have already done the best you can at an assignment and it would already earn a very good grade. Instead you should now concentrate on improving an assignment which covers an area where you know you are weak, because this will boost your overall grade(s). You will learn more about this in Step 3 – Use your time wisely.

PLUSPOINTS

+ Your learning is assessed in a variety of ways, such as by assignments, projects and case studies. You will need to be able to research effectively, manage your own time and work well with other people to succeed.

+ You need to demonstrate specific knowledge and skills to achieve the learning outcomes set by Edexcel. You need to demonstrate you can meet all the learning outcomes to pass a unit.

+ Higher-level skills are required for higher grades. The grading criteria for Pass, Merit and Distinction grades are set out in a grading grid for the unit.

+ The assessment grades of Pass, Merit and Distinction convert to unit points. The total number of unit points you receive during the course determines your final overall grade(s) and the UCAS points you have earned.

+ Working effectively from the beginning maximises your chances of achieving a good qualification grade. Understanding grade boundaries enables you to get the best final grade(s) possible.

ACTION POINTS

✓ Find the learning outcomes for the units you are currently studying. Your tutor may have given you these already, or you can find them in the specification for your course that you already accessed at www.edexcel.org.uk.

✓ Look at the grading grid for the units and identify the way the evidence required changes to achieve the higher grades. Don't worry if there are some words that you do not understand – these are explained in more detail on page 32 of this guide.

✓ If you are still unsure how the unit points system works, ask your tutor to explain it to you.

✓ Check out the number of UCAS points you would need for any course or university in which you are interested.

✓ Keep a record of the unit points you earn throughout your course and check regularly how this is affecting your overall grade(s), based on the grade boundaries for your qualification. Your tutor will give you this information or you can check it yourself in the specification for your course on the Edexcel website.

STEP THREE

USE YOUR TIME WISELY

Most students on a BTEC National course are trying to combine their course commitments with a number of others – such as a job (either full or part-time) and family responsibilities. In addition, they still want time to meet with friends, enjoy a social life and keep up hobbies and interests that they have.

Starting the course doesn't mean that you have to hide away for months if you want to do well. It does mean that you have to use your time wisely if you want to do well, stay sane and keep a balance in your life.

You will only do this if you make time work for you, rather than against you, by taking control. This means that you decide what you are doing, when you are doing it and work purposefully; rather than simply reacting to problems or panicking madly because you've yet another deadline staring you in the face.

This becomes even more important as your course progresses because your workload is likely to increase, particularly towards the end of a term. In the early days you may be beautifully organised and able to cope easily. Then you may find you have several tasks to complete simultaneously as well as some research to start. Then you get two assignments in the same week from different tutors – as well as having a presentation to prepare. Then another assignment is scheduled for the following week – and so on. This is not because your tutors are being deliberately difficult. Indeed, most will try to schedule your assignments to avoid such clashes. The problem, of course, is that none of your tutors can assess your abilities until you have learned something – so if several units start and end at the same time it is highly likely there will be some overlap between your assignments.

To cope when the going gets tough, without collapsing into an exhausted heap, you need to learn a few time management skills.

Use your time wisely

- **Pinpoint where your time goes at the moment** Time is like money – it's usually difficult to work out where it all went! Work out how much time you currently spend at college, at work, at home and on social activities. Check, too, how much time you waste each week – and why this happens. Are you disorganised or do you easily get distracted? Then identify commitments that are vital and those that are optional so that you know where you can find time if you need to.

- **Plan when and where to work** It is unrealistic not to expect to do quite a lot of work for your course in your own time. It is also better to work regularly, and in relatively short bursts, than to work just once or twice a week for very long stretches. In addition to deciding when to work, and for how long, you also need to think about when and where to work. If you are a lark, you will work better early in the day; if you are an owl, you will be at your best later on. Whatever time you work, you need somewhere quiet so that you can concentrate and with space for books and other resources you need. If the words 'quiet oasis' and 'your house' are totally incompatible at any time of the day or night then check out the opening hours of your local and college library so that you have an escape route if you need it. If you are trying to combine studying with parental responsibilities it is sensible to factor in your children's commitments – and work around their bedtimes too! Store up favours, too, from friends and grandparents that you can call in if you get desperate for extra time when an assignment deadline is looming.

- **Schedule your commitments** Keep a diary or (even better) a wall chart and write down every appointment you make or task you are given. It is useful to use a colour code to differentiate between personal and work or course commitments. You may also want to enter assignment review dates with your tutor in one colour and final deadline dates in another. Keep your diary or chart up-to-date by adding any new dates promptly every time you receive another task or assignment or whenever you make any other arrangements. Keep checking ahead so that you always have prior warning when important dates are looming. This stops you from planning a heavy social week when you will be at your busiest at work or college and from arranging a dental appointment on the morning when you and your team are scheduled to give an important presentation!

- **Prioritise your work** This means doing the most important and urgent task first, rather than the one you like the most! Normally this will be the task or assignment with the nearest deadline. There are two exceptions. Sometimes you may need to send off for information and allow time for it to arrive. It is therefore sensible to do this first so that you are not held up later. The second is when you have to take account of other people's schedules – because you are working in a team or are arranging to interview someone, for example. In this case you will have to arrange your schedule around their needs, not just your own.

- **Set sensible timescales** Trying to do work at the last minute or in a rush is never satisfactory, so it is wise always to allocate more time than you think you will need, never less. Remember, too, to include all the stages of a complex task or assignment, such as researching the information, deciding what to use, creating a first draft, checking it and making improvements and printing it out. If you are planning to do any of your work in a central facility always allow extra time and try to start work early. If you arrive at the last minute you may find every computer and printer is fully utilised until closing time.

- **Learn self-discipline!** This means not putting things off (procrastinating!) because you don't know where to start or don't feel in the mood. Unless you are ill, you have to find some way of persuading yourself to work. One way is to bribe yourself. Make a start and promise yourself that if you work productively for 30 minutes then you deserve a small reward. After 30 minutes you may have become more engrossed and want to keep going a little longer. Otherwise at least you have made a start, so it's easier to come back and do more later. It doesn't matter whether you have research to do, an assignment to write up, a coaching session to plan, or lines to learn, you need to be self-disciplined.

- **Take regular breaks and keep your life in balance** Don't go to the opposite extreme and work for hours on end. Take regular breaks to give yourself a rest and a change of activity. You need to recharge your batteries! Similarly, don't cancel every social arrangement so that you can work 24/7. Whilst this may be occasionally necessary if you have several deadlines looming simultaneously, it should only be a last resort. If you find yourself doing this regularly then go back to the beginning of this section and see where your time-management planning is going wrong.

PLUSPOINTS

+ Being in control of your time enables you to balance your commitments according to their importance and allows you not let to anyone down – including yourself.

+ Controlling time involves knowing how you spend (and waste!) your time now, planning when best to do work, scheduling your commitments and setting sensible timescales for work to be done.

+ Knowing how to prioritise means that you will schedule work effectively according to its urgency and importance but this also requires self-discipline. You have to follow the schedule you have set for yourself!

+ Managing time and focusing on the task at hand means you will do better work and be less stressed, because you are not having to react to problems or crises. You can also find the time to include regular breaks and leisure activities in your schedule.

ACTION POINTS

✓ Find out how many assignments you can expect to receive this term and when you can expect to receive these. Enter this information into your student diary or onto a planner you can refer to regularly.

✓ Update your diary and/or planner with other commitments that you have this term – both work/college-related and social. Identify any potential clashes and decide the best action to take to solve the problem.

✓ Identify your own best time and place to work quietly and effectively.

✓ Displacement activities are things we do to put off starting a job we don't want to do – such as sending texts, watching TV, checking emails etc. Identify yours so that you know when you're doing them!

STEP FOUR

UTILISE ALL YOUR RESOURCES

Your resources are all the things that can help you to achieve your qualification. They can therefore be as wide-ranging as your favourite website and your **study buddy** (see page 15) who collects handouts for you if you miss a class.

Your college will provide the essential resources for your course, such as a library with a wide range of books and electronic reference sources, learning resource centre(s), the computer network and Internet access. Other basic resources you will be expected to provide yourself, such as file folders and paper. The policy on textbooks varies from one college to another, but on most courses today students are expected to buy their own. If you look after yours carefully, then you have the option to sell it on to someone else afterwards and recoup some of your money. If you scribble all over it, leave it on the floor and then tread on it, turn back pages and rapidly turn it into a dog-eared, misshapen version of its former self then you miss out on this opportunity.

Unfortunately students often squander other opportunities to utilise resources in the best way – usually because they don't think about them very much, if at all. To help, below is a list of the resources you should consider important – with a few tips on how to get the best out of them.

- **Course information** This includes your course specification, this Study Guide and all the other information relating to your BTEC National which you can find on the Edexcel website. Add to this all the information given to you at college relating to your course, including term dates, assignment dates and, of course, your timetable. This should not be 'dead' information that you glance at once and then discard or ignore. Rather it is important reference material that you need to store somewhere obvious, so that you can look at it whenever you have a query or need to clarify something quickly.

- **Course materials** In this group is your textbook (if there is one), the handouts you are given as well as print-outs and notes you make yourself. File handouts the moment you are given them and put them into an A4 folder bought for the purpose. You will need one for each unit you study. Some students prefer lever-arch files but these are more bulky so more difficult to carry around all day. Unless you have a locker at college it can be easier to keep a lever arch file at home for permanent storage of past handouts and notes for a unit and carry an A4 folder with you which contains current topic information. Filing handouts and print-outs promptly means they don't get lost. They are also less likely to get crumpled, torn or tatty becoming virtually unreadable. Unless you have a private and extensive source of income then this is even more important if you have to pay for every print-out you take in your college resource centre. If you are following a course such as Art and Design, then there will be all your art materials and the pieces you produce. You must look after these with great care.

- **Other stationery items** Having your own pens, pencils, notepad, punch, stapler and sets of dividers is essential. Nothing irritates tutors more than watching one punch circulate around a group – except, perhaps, the student who trudges into class with nothing to write on or with. Your dividers should be clearly labelled to help you store and find notes, print-outs and handouts fast. Similarly, your notes should be clearly headed and dated. If you are writing notes up from your own research then you will have to include your source. Researching information is explained in Step 6 – Sharpen your skills.

- **Equipment and facilities** These include your college library and resource centres, the college computer network and other college equipment you can use, such as laptop computers, photocopiers and presentation equipment. Much of this may be freely available; others – such as using the photocopier in the college library or the printers in a resource centre – may cost you money. Many useful resources will be electronic, such as DVDs or electronic journals and databases. At home you may have your own computer with Internet access to count as a resource. Finally, include any specialist equipment and facilities available for your particular course that you use at college or have at home.

Utilise all your resources

All centralised college resources and facilities are invaluable if you know how to use them – but can be baffling when you don't. Your induction should have included how to use the library, resource centre(s) and computer network. You should also have been informed of the policy on using IT equipment which determines what you can and can't do when you are using college computers. If, by any chance, you missed this then go and check it out for yourself. Library and resource centre staff will be only too pleased to give you helpful advice – especially if you pick a quiet time to call in. You can also find out about the allowable ways to transfer data between your college computer and your home computer if your options are limited because of IT security.

Having a study buddy is a good idea

■ **People** You are surrounded by people who are valuable resources: your tutor(s), specialist staff at college, your employer and work colleagues, your relatives and any friends who have particular skills or who work in the same area you are studying. Other members of your class are also useful resources – although they may not always seem like it! Use them, for example, to discuss topics out of class time. A good debate between a group of students can often raise and clarify issues that there may not be time to discuss fully in class. Having a study buddy is another good idea – you get/make notes for them when they are away and vice versa. That way you don't miss anything.

If you want information or help from someone, especially anyone outside your immediate circle, then remember to get the basics right! Approach them courteously, do your homework first so that you are well-prepared and remember that you are asking for assistance – not trying to get them to do the work for you! If someone has agreed to allow you to interview them as part of your research for an assignment or project then good preparations will be vital, as you will see in Step 6 – Sharpen your Skills (see page 22).

One word of warning: be wary about using information from friends or relatives who have done a similar or earlier course. First, the slant of the material they were given may be different. It may also be out-of-date. And *never* copy anyone else's written assignments. This is **plagiarism** – a deadly sin in the educational world. You can read more about this in Step 5 – Understand your assessment (see page 16).

■ **You!** You have the ability to be your own best resource or your own worst enemy! The difference depends upon your work skills, your personal skills and your attitude to your course and other people. You have already seen how to use time wisely. Throughout this guide you will find out how to sharpen and improve other work and personal skills and how to get the most out of your course – but it is up to you to read it and apply your new-found knowledge! This is why attributes like a positive attitude, an enquiring mind and the ability to focus on what is important all have a major impact on your final result.

PLUSPOINTS

+ Resources help you to achieve your qualification. You will squander these unwittingly if you don't know what they are or how to use them properly.

+ Course information needs to be stored safely for future reference: course materials need to be filed promptly and accurately so that you can find them quickly.

+ You need your own set of key stationery items; you also need to know how to use any central facilities or resources such as the library, learning resource centres and your computer network.

+ People are often a key resource – school or college staff, work colleagues, members of your class, people who are experts in their field.

+ You can be your own best resource! Develop the skills you need to be able to work quickly and accurately and to get the most out of other people who can help you.

ACTION POINTS

✓ Under the same headings as in this section, list all the resources you need for your course and tick off those you currently have. Then decide how and when you can obtain anything vital that you lack.

✓ Check that you know how to access and use all the shared resources to which you have access at school or college.

✓ Pair up with someone on your course as a study buddy – and don't let them down!

✓ Test your own storage systems. How fast can you find notes or print-outs you made yesterday/last week/last month – and what condition are they in?

✓ Find out the IT policy at your school or college and make sure you abide by it.

STEP FIVE

UNDERSTAND YOUR ASSESSMENT

The key to doing really, really well on any BTEC National course is to understand exactly what you are expected to do in your assignments – and then to do it! It really is as simple as that. So why is it that some people go wrong?

Obviously you may worry that an assignment may be so difficult that it is beyond you. Actually this is highly unlikely to happen because all your assignments are based on topics you will have already covered thoroughly in class. Therefore, if you have attended regularly – and have clarified any queries or worries you have either in class or during your tutorials – this shouldn't happen. If you have had an unavoidable lengthy absence then you may need to review your progress with your tutor and decide how best to cope with the situation. Otherwise, you should note that the main problems with assignments are usually due to far more mundane pitfalls – such as:

✗ not reading the instructions or the assignment brief properly

✗ not understanding what you are supposed to do

✗ only doing part of the task or answering part of a question

✗ skimping the preparation, the research or the whole thing

✗ not communicating your ideas clearly

✗ guessing answers rather than researching properly

✗ padding out answers with irrelevant information

✗ leaving the work until the last minute and then doing it in a rush

✗ ignoring advice and feedback your tutor has given you.

You can avoid all of these traps by following the guidelines below so that you know exactly what you are doing, prepare well and produce your best work.

The assignment 'brief'

The word 'brief' is just another way of saying 'instructions'. Often, though, a 'brief' (despite its name!) may be rather longer. The brief sets the context for the work, defines what evidence you will need to produce and matches the grading criteria to the tasks. It will also give you a schedule for completing the tasks. For example, a brief may include details of a case study you have to read; research you have to carry out or a task you have to do, as well as questions you have to answer. Or it may give you details about a project or group presentation you have to prepare. The type of assignments you receive will depend partly upon the vocational area you are studying, but you can expect some to be in the form of written assignments. Others are likely to be more practical or project-based, especially if you are doing a very practical subject such as Art and Design, Performing Arts or Sport. You may also be assessed in the workplace. For example, this is a course requirement if you are studying Children's Care, Learning and Development.

The assignment brief may also include the **learning outcomes** to which it relates. These tell you the purpose of the assessment and the knowledge you need to demonstrate to obtain a required grade. If your brief doesn't list the learning outcomes, then you should check this information against the unit specification to see the exact knowledge you need to demonstrate.

The grade(s) you can obtain will also be stated on the assignment brief. Sometimes an assignment will focus on just one grade. Others may give you the opportunity to develop or extend your work to progress to a higher grade. This is often dependent upon submitting acceptable work at the previous grade first. You will see examples of this in the Marked Assignment section of this Study Guide on page 119.

The brief will also tell you if you have to do part of the work as a member of a group. In this case, you must identify your own contribution. You may also be expected to take part in a **peer review**, where you all give feedback on the contribution of one another. Remember that you should do this as objectively and professionally as possible – not just praise everyone madly in the hope that they will do the same for you! In any assignment where there is a group contribution, there is virtually always an individual component, so that your individual grade can be assessed accurately.

Finally, your assignment brief should state the final deadline for handing in the work as well as any interim review dates when you can discuss your progress and ideas with your tutor. These are very important dates indeed and should be entered immediately into your diary or planner. You should schedule your work around these dates so that you have made a start by the first date. This will then enable you to note any queries or significant issues you want to discuss. Otherwise you will waste a valuable opportunity to obtain useful feedback on your progress. Remember, too, to take a notebook to any review meetings so that you can write down the guidance you are given.

Your school or college rules and regulations

Your school or college will have a number of policies and guidelines about assignments and assessment. These will deal with issues such as:

- The procedure you must follow if you have a serious personal problem so cannot meet the deadline date and need an extension.

- Any penalties for missing a deadline date without any good reason.

- The penalties for copying someone else's work (**plagiarism**). These will be severe so make sure that you never share your work (including your CDs) with anyone else and don't ask to borrow theirs.

- The procedure to follow if you are unhappy with the final grade you receive.

Even though it is unlikely that you will ever need to use any of these policies, it is sensible to know they exist, and what they say, just as a safeguard.

Understanding the question or task

There are two aspects to a question or task that need attention. The first are the *command words*, which are explained below. The second are the *presentation instructions*, so that if you are asked to produce a table or graph or report then you do exactly that – and don't write a list or an essay instead!

Command words are used to specify how a question must be answered, eg 'explain', 'describe', 'analyse', 'evaluate'. These words relate to the type of answer required. So whereas you may be asked to 'describe' something at Pass level, you will need to do more (such as 'analyse' or 'evaluate') to achieve Merit or Distinction grade.

Many students fail to get a higher grade because they do not realise the difference between these words. They simply don't know *how* to analyse or evaluate, so give an explanation instead. Just adding to a list or giving a few more details will never give you a higher grade – instead you need to change your whole approach to the answer.

The **grading grid** for each unit of your course gives you the command words, so that you can find out exactly what you have to do in each unit, to obtain a Pass, Merit and Distinction. The following charts show you what is usually required when you see a particular command word. You can use this, and the marked assignments on pages 119–142, to see the difference between the types of answers required for each grade. (The assignments your centre gives you will be specially written to ensure you have the opportunity to achieve all the possible grades.) Remember, though, that these are just examples to guide you. The exact response will often depend upon the way a question is worded, so if you have any doubts at all check with your tutor before you start work.

There are two other important points to note:

- Sometimes the same command word may be repeated for different grades – such as 'create' or 'explain'. In this case the *complexity* or *range* of the task itself increases at the higher grades – as you will see if you read the grading grid for the unit.

- Command words can also vary depending upon your vocational area. If you are studying Performing Arts or Art and Design you will probably find several command words that an Engineer or IT Practitioner would not – and vice versa!

To obtain a Pass grade

To achieve this grade you must usually demonstrate that you understand the important facts relating to a topic and can state these clearly and concisely.

Command word	What this means
Create (or produce)	Make, invent or construct an item.
Describe	Give a clear, straightforward description that includes all the main points and links these together logically.
Define	Clearly explain what a particular term means and give an example, if appropriate, to show what you mean.
Explain . . . how/why	Set out in detail the meaning of something, with reasons. It is often helpful to give an example of what you mean. Start with the topic then give the 'how' or 'why'.
Identify	Distinguish and state the main features or basic facts relating to a topic.
Interpret	Define or explain the meaning of something.
Illustrate	Give examples to show what you mean.
List	Provide the information required in a list rather than in continuous writing.
Outline	Write a clear description that includes all the main points but avoid going into too much detail.
Plan (or devise)	Work out and explain how you would carry out a task or activity.
Select (and present) information	Identify relevant information to support the argument you are making and communicate this in an appropriate way.
State	Write a clear and full account.
Undertake	Carry out a specific activity.
Examples: **Identify** the main features on a digital camera. **Describe** your usual lifestyle. **Outline** the steps to take to carry out research for an assignment.	

19

To obtain a Merit grade

To obtain this grade you must prove that you can apply your knowledge in a specific way.

Command word	What this means
Analyse	Identify separate factors, say how they are related and how each one relates to the topic.
Classify	Sort your information into appropriate categories before presenting or explaining it.
Compare and contrast	Identify the main factors that apply in two or more situations and explain the similarities and differences or advantages and disadvantages.
Demonstrate	Provide several relevant examples or appropriate evidence which support the arguments you are making. In some vocational areas this may also mean giving a practical performance.
Discuss	Provide a thoughtful and logical argument to support the case you are making.
Explain (in detail)	Provide details and give reasons and/or evidence to clearly support the argument you are making.
Implement	Put into practice or operation. You may also have to interpret or justify the effect or result.
Interpret	Understand and explain an effect or result.
Justify	Give appropriate reasons to support your opinion or views and show how you arrived at these conclusions.
Relate/report	Give a full account of, with reasons.
Research	Carry out a full investigation.
Specify	Provide full details and descriptions of selected items or activities.

Examples:

Compare and contrast the performance of two different digital cameras.
Justify your usual lifestyle.
Explain in detail the steps to take to research an assignment.

To obtain a Distinction grade

To obtain this grade you must prove that you can make a reasoned judgement based on appropriate evidence.

Command word	What this means
Analyse	Identify the key factors, show how they are linked and explain the importance and relevance of each.
Assess	Give careful consideration to all the factors or events that apply and identify which are the most important and relevant with reasons for your views.
Comprehensively explain	Give a very detailed explanation that covers all the relevant points and give reasons for your views or actions.
Comment critically	Give your view after you have considered all the evidence, particularly the importance of both the relevant positive and negative aspects.
Evaluate	Review the information and then bring it together to form a conclusion. Give evidence to support each of your views or statements.
Evaluate critically	Review the information to decide the degree to which something is true, important or valuable. Then assess possible alternatives taking into account their strengths and weaknesses if they were applied instead. Then give a precise and detailed account to explain your opinion.
Summarise	Identify and review the main, relevant factors and/or arguments so that these are explained in a clear and concise manner.
Examples: **Assess** ten features commonly found on a digital camera. **Evaluate critically** your usual lifestyle. **Analyse** your own ability to carry out effective research for an assignment.	

Responding positively

This is often the most important attribute of all! If you believe that assignments give you the opportunity to demonstrate what you know and how you can apply it *and* positively respond to the challenge by being determined to give it your best shot, then you will do far better than someone who is defeated before they start.

It obviously helps, too, if you are well organised and have confidence in your own abilities – which is what the next section is all about!

PLUSPOINTS

+ Many mistakes in assignments are through errors that can easily be avoided such as not reading the instructions properly or doing only part of the task that was set!

+ Always read the assignment brief very carefully indeed. Check that you understand exactly what you have to do and the learning outcomes you must demonstrate.

+ Make a note of the deadline for an assignment and any interim review dates on your planner. Schedule work around these dates so that you can make the most of reviews with your tutor.

+ Make sure you know about school or college policies relating to assessment, such as how to obtain an extension or query a final grade.

+ For every assignment, make sure you understand the command words, which tell you how to answer the question, and the presentation instructions, which say what you must produce.

+ Command words are shown in the grading grid for each unit of your qualification. Expect command words and/or the complexity of a task to be different at higher grades, because you have to demonstrate higher-level skills.

ACTION POINTS

✓ Discuss with your tutor the format (style) of assignments you are likely to receive on your course, eg assignments, projects, or practical work where you are observed.

✓ Check the format of the assignments in the Marked Assignments section of this book. Look at the type of work students did to gain a Pass and then look at the difference in the Merit answers. Read the tutor's comments carefully and ask your own tutor if there is anything you do not understand.

✓ Check out all the policies and guidelines at your school or college that relate to assessment and make sure you understand them.

✓ Check out the grading grid for the units you are currently studying and identify the command words for each grade. Then check you understand what they mean using the explanations above. If there are any words that are not included, ask your tutor to explain the meanings and what you would be required to do.

STEP SIX

SHARPEN YOUR SKILLS

To do your best in any assignment you need a number of skills. Some of these may be vocationally specific, or professional, skills that you are learning as part of your course – such as acting or dancing if you are taking a Performing Arts course or, perhaps, football if you are following a Sports course. Others, though, are broader skills that will help you to do well in assignments no matter what subjects or topics you are studying – such as communicating clearly and cooperating with others.

Some of these skills you will have already and in some areas you may be extremely proficient. Knowing where your weaknesses lie, though, and doing something about them has many benefits. You will work more quickly, more accurately *and* have increased confidence in your own abilities. As an extra bonus, all these skills also make you more effective at work – so there really is no excuse for not giving yourself a quick skills check and then remedying any problem areas.

This section contains hints and tips to help you check out and improve each of the following areas:

- Your numeracy skills
- Keyboarding and document preparation
- Your IT skills
- Your written communication skills
- Working with others
- Researching information
- Making a presentation

Your numeracy skills

Some people have the idea that they can ignore numeracy because this skill isn't relevant to their vocational area – such as Art and Design or Children's Care, Learning and Development. If this is how you think then you are wrong! Numeracy is a life skill that everyone needs, so if you can't carry out basic calculations accurately then you will have problems, often when you least expect them.

Fortunately there are several things you can do to remedy this situation:

- Practise basic calculations in your head and then check them on a calculator.
- Ask your tutor if there are any essential calculations which give you difficulties.
- Use your onscreen calculator (or a spreadsheet package) to do calculations for you when you are using your computer.
- Try your hand at Sudoku puzzles – either on paper or by using a software package or online at sites such as www.websudoku.com/.
- Investigate puzzle sites and brain training software, such as http://school.discovery.com/brainboosters/ and Dr Kawashima's Brain Training by Nintendo.
- Check out online sites such as www.bbc.co.uk/skillswise/ and www.bbc.co.uk/schools/ks3bitesize/maths/number/index.shtml to improve your skills.

Numeracy is a life skill

Keyboarding and document preparation

- Think seriously about learning to touch type to save hours of time! Your school or college may have a workshop you can join or you can learn online such as by downloading a free program at www.sense-lang.org/typing/ or practising on sites such as www.computerlab.kids.new.net/keyboarding.htm.
- Obtain correct examples of document formats you will have to use, such as a report or summary. Your tutor may provide you with these or you can find examples in many communication textbooks.
- Proof-read work you produce on a computer *carefully*. Remember that your spell checker will not pick up every mistake you make, such as a mistyped word that makes another word (eg form/from; sheer/shear)

23

and grammar checkers, too, are not without their problems! This means you still have to read your work through yourself. If possible, let your work go 'cold' before you do this so that you read it afresh and don't make assumptions about what you have written. Then read word by word to make sure it still makes sense and there are no silly mistakes, such as missing or duplicated words.

■ Make sure your work looks professional by using an appropriate typeface and font size as well as suitable margins.

■ Print out your work carefully and store it neatly, so it looks in pristine condition when you hand it in.

Your IT skills

■ Check that you can use the main features of all the software packages that you will need to produce your assignments, such as Word, Excel and PowerPoint.

■ Adopt a good search engine, such as Google, and learn to use it properly. Many have online tutorials such as www.googleguide.com.

■ Develop your IT skills to enable you to enhance your assignments appropriately. For example, this may include learning how to import and export text and artwork from one package to another; taking digital photographs and inserting them into your work and/or creating drawings or diagrams by using appropriate software for your course.

Your written communication skills

A poor vocabulary will reduce your ability to explain yourself clearly; work peppered with spelling or punctuation errors looks unprofessional.

■ Read more. This introduces you to new words and familiarises you over and over again with the correct way to spell words.

■ Look up words you don't understand in a dictionary and then try to use them yourself in conversation.

■ Use the Thesaurus in Word to find alternatives to words you find yourself regularly repeating, to add variety to your work.

■ *Never* use words you don't understand in the hope that they sound impressive!

■ Do crosswords to improve your word power and spelling.

■ Resolve to master punctuation – especially apostrophes – either by using an online programme or working your way through the relevant section of a communication textbook that you like.

■ Check out online sites such as www.bbc.co.uk/skillswise/ and www.bbc.co.uk/schools/gcsebitesize/english/ as well as puzzle sites with communication questions such as http://school.discovery.com/brainboosters/.

Working with others

In your private life you can choose who you want to be with and how you respond to them. At work you cannot do that – you are paid to be professional and this means working alongside a wide variety of people, some of whom you may like and some of whom you may not!

The same applies at school or college. By the time you have reached BTEC National level you will be expected to have outgrown wanting to work with your best friends on every project! You may not be very keen on everyone who is in the same team as you, but – at the very least – you can be pleasant, cooperative and helpful. In a large group this isn't normally too difficult. You may find it much harder if you have to partner someone who has very different ideas and ways of working to you.

In this case it may help if you:

- Realise that everyone is different and that your ways of working may not always be the best!
- Are prepared to listen and contribute to a discussion (positively) in equal amounts. Make sure, too, that you encourage the quiet members of the group to speak up by asking them what their views are. The ability to draw other people into the discussion is an important and valuable skill to learn.
- Write down what you have said you will do, so that you don't forget anything.
- Are prepared to do your fair share of the work.
- Discuss options and alternatives with people – don't give them orders or meekly accept instructions and then resent it afterwards.
- Don't expect other people to do what you wouldn't be prepared to do.
- Are sensitive to other people's feelings and remember that they may have personal problems or issues that affect their behaviour.
- *Always* keep your promises and never let anyone down when they are depending upon you.
- Don't flounce around or lose your temper if things get tough. Instead take a break while you cool down. Then sit down and discuss the issues that are annoying you.
- Help other people to reach a compromise when necessary, by acting as peacemaker.

Researching information

Poor researchers either cannot find what they want or find too much – and then drown in a pile of papers. If you find yourself drifting aimlessly around a library when you want information or printing out dozens of pages for no apparent purpose, then this section is for you!

- Always check *exactly* what it is you need to find and how much detail is needed. Write down a few key words to keep yourself focused.
- Discipline yourself to ignore anything that is irrelevant – from books with interesting titles to websites which sound tempting but have little to do with your topic or key words.
- Remember that you could theoretically research information forever! So at some time you have to call a halt. Learning when to do this is another skill, but you can learn this by writing out a schedule which clearly states when you have to stop looking and start sorting out your information and writing about it!

- In a library, check you know how the books are stored and what other types of media are available. If you can't find what you are looking for then ask the librarian for help. Checking the index in a book is the quickest way to find out whether it contains information related to your key words. Put it back if it doesn't or if you can't understand it. If you find three or four books and/or journals that contain what you need then that is usually enough.

- Online use a good search engine and use the summary of the search results to check out the best sites. Force yourself to check out sites beyond page one of the search results! When you enter a site investigate it carefully – use the site map if necessary. It isn't always easy to find exactly what you want. Bookmark sites you find helpful and will want to use again and only take print-outs when the information is closely related to your key words.

- Talk to people who can help you (see also Step 4 – Utilise all your resources) and prepare in advance by thinking about the best questions to ask. Always explain why you want the information and don't expect anyone to tell you anything that is confidential or sensitive – such as personal information or financial details. Always write clear notes so that you remember what you have been told, by whom and when. If you are wise you will also note down their contact details so that you can contact them again if you think of anything later. If you remember to be courteous and thank them for their help, this shouldn't be a problem.

- Store all your precious information carefully and neatly in a labelled folder so that you can find it easily. Then, when you are ready to start work, reread it and extract that which is most closely related to your key words and the task you are doing.

- Make sure you state the source of all the information you quote by including the name of the author or the web address, either in the text or as part of a bibliography at the end. Your school or college will have a help sheet which will tell you exactly how to do this.

Making a presentation

This involves several skills – which is why it is such a popular way of finding out what students can do! It will test your ability to work in a team, speak in public and use IT (normally PowerPoint) – as well as your nerves. It is therefore excellent practice for many of the tasks you will have to do when you are at work – from attending an interview to talking to an important client.

You will be less nervous if you have prepared well and have rehearsed your role beforehand. You will produce a better, more professional presentation if you take note of the following points.

- If you are working as a team, work out everyone's strengths and weaknesses and divide up the work (fairly) taking these into account. Work out, too, how long each person should speak and who would be the best as the 'leader' who introduces each person and then summarises everything at the end.

PLUSPOINTS

+ Poor numeracy skills can let you down in your assignments and at work. Work at improving these if you regularly struggle with even simple calculations.

+ Good keyboarding, document production and IT skills can save you hours of time and mean that your work is of a far more professional standard. Improve any of these areas which are letting you down.

+ Your written communication skills will be tested in many assignments. Work at improving areas of weakness, such as spelling, punctuation or vocabulary.

+ You will be expected to work cooperatively with other people both at work and during many assignments. Be sensitive to other people's feelings, not just your own, and always be prepared to do your fair share of the work and help other people when you can.

+ To research effectively you need to know exactly what you are trying to find and where to look. This means understanding how reference media is stored in your library as well as how to search online. Good organisation skills also help so that you store important information carefully and can find it later. And never forget to include your sources in a bibliography.

+ Making a presentation requires several skills and may be nerve-racking at first. You will reduce your problems if you prepare well, are not too ambitious and have several run-throughs beforehand. Remember to speak clearly and a little more slowly than normal and smile from time to time!

ACTION POINTS

✓ Test both your numeracy and literacy skills at http://www.move-on.org.uk/testyourskills.asp# to check your current level. You don't need to register on the site if you click to do the 'mini-test' instead. If either need improvement, get help at http://www.bbc.co.uk/keyskills/it/1.shtml.

✓ Do the following two tasks with a partner to jerk your brain into action!

  – Each write down 36 simple calculations in a list, eg 8 x 6, 19 – 8, 14 + 6. Then exchange lists. See who can answer the most correctly in the shortest time.

  – Each write down 30 short random words (no more than 8 letters), eg cave, table, happily. Exchange lists. You each have three minutes to try to remember as many words as possible. Then hand back the list and write down all those you can recall. See who can remember the most.

✓ Assess your own keyboarding, proof-reading, document production, written communication and IT skills. Then find out if your tutors agree with you!

✓ List ten traits in other people that drive you mad. Then, for each one, suggest what you could do to cope with the problem (or solve it) rather than make a fuss. Compare your ideas with other members of your group.

✓ Take a note of all feedback you receive from your tutors, especially in relation to working with other people, researching and giving presentations. In each case focus on their suggestions and ideas so that you continually improve your skills throughout the course.

27

■ Don't be over-ambitious. Take account of your time-scale, resources and the skills of the team. Remember that a simple, clear presentation is often more professional than an over-elaborate or complicated one where half the visual aids don't work properly!

■ If you are using PowerPoint try to avoid preparing every slide with bullet points! For variety, include some artwork and vary the designs. Remember that you should *never* just read your slides to the audience! Instead prepare notes that you can print out that will enable you to enhance and extend what the audience is reading.

- Your preparations should also include checking the venue and time; deciding what to wear and getting it ready; preparing, checking and printing any handouts; deciding what questions might be asked and how to answer these.

- Have several run-throughs beforehand and check your timings. Check, too, that you can be heard clearly. This means lifting up your head and 'speaking' to the back of the room a little more slowly and loudly than you normally do.

- On the day, arrive in plenty of time so that you aren't rushed or stressed. Remember that taking deep breaths helps to calm your nerves.

- Start by introducing yourself clearly and smile at the audience. If it helps, find a friendly face and pretend you are just talking to that person.

- Answer any questions honestly and don't exaggerate, guess or waffle. If you don't know the answer then say so!

- If you are giving the presentation in a team, help out someone else who is struggling with a question if you know the answer.

- Don't get annoyed or upset if you get any negative feedback afterwards. Instead take note so that you can concentrate on improving your own performance next time. And don't focus on one or two criticisms and ignore all the praise you received! Building on the good and minimising the bad is how everyone improves in life!

28

STEP SEVEN

MAXIMISE YOUR OPPORTUNITIES AND MANAGE YOUR PROBLEMS

Like most things in life, you may have a few ups and downs on your course – particularly if you are studying over quite a long time, such as one or two years. Sometimes everything will be marvellous – you are enjoying all the units, you are up-to-date with your work, you are finding the subjects interesting and having no problems with any of your research tasks. At other times you may struggle a little more. You may find one or two topics rather tedious, or there may be distractions or worries in your personal life that you have to cope with. You may struggle to concentrate on the work and do your best.

Rather than just suffering in silence or gritting your teeth if things go a bit awry it is sensible if you have an action plan to help you cope. Equally, rather than just accepting good opportunities for additional experiences or learning, it is also wise to plan how to make the best of these. This section will show you how to do this.

Making the most of your opportunities

The following are examples of opportunities to find out more about information relevant to your course or to try putting some of your skills into practice.

- **External visits** You may go out of college on visits to different places or organisations. These are not days off – there is a reason for making each trip. Prepare in advance by reading around relevant topics and make notes of useful information whilst you are there. Then write (or type) it up neatly as soon as you can and file it where you can find it again!

- **Visiting speakers** Again, people are asked to talk to your group for a purpose. You are likely to be asked to contribute towards questions that may be asked – which may be submitted in advance so that the speaker is clear on the topics you are studying. Think carefully about information that you would find helpful so that you can ask one or two relevant and useful questions. Take notes whilst the speaker is addressing your group, unless someone is recording the session. Be prepared to thank the speaker on behalf of your group if you are asked to do so.

- **Professional contacts** These will be the people you meet on work experience doing the real job that one day you hope to do. Make the most of meeting these people to find out about the vocational area of your choice.

- **Work experience** If you need to undertake practical work for any particular units of your BTEC National qualification, and if you are studying full-time, then your tutor will organise a work experience placement for you and talk to you about the evidence you need to obtain. You may also be issued with a special log book or diary in which to record your experiences. Before you start your placement, check that you are clear about all the details, such as the time you will start and leave, the name of your supervisor, what you should wear and what you should do if you are ill during the placement and cannot attend. Read and reread the units to which your evidence will apply and make sure you understand the grading criteria and what you need to obtain. Then make a note of appropriate headings to record your information. Try to make time to write up your notes, log book and/or diary every night, whilst your experiences are fresh in your mind.

- **In your own workplace** You may be studying your BTEC National qualification on a part-time basis and also have a full-time job in the same vocational area. Or you may be studying full-time and have a part-time job just to earn some money. In either case you should be alert to opportunities to find out more about topics that relate to your workplace, no matter how generally. For example, many BTEC courses include topics such as health and safety, teamwork, dealing with customers, IT security and communications – to name but a few. All these are topics that your employer will have had to address and finding out more about these will broaden your knowledge and help to give more depth to your assignment responses.

- **Television programmes, newspapers, Podcasts and other information sources** No matter what vocational area you are studying, the media are likely to be an invaluable source of information. You should be alert to any news bulletins that relate to your studies as well as relevant information in more topical television programmes. For example, if you are studying Art and Design then you should make a particular effort to watch the *Culture Show* as well as programmes on artists, exhibitions

29

or other topics of interest. Business students should find inspiration by watching *Dragons Den*, *The Apprentice* and the *Money Programme* and Travel and Tourism students should watch holiday, travel and adventure programmes. If you are studying Media, Music and Performing Arts then you are spoiled for choice! Whatever your vocational choice, there will be television and radio programmes of special interest to you.

Remember that you can record television programmes to watch later if you prefer, and check out newspaper headlines online and from sites such as BBC news. The same applies to Podcasts. Of course, to know which information is relevant means that you must be familiar with the content of all the units you are studying, so it is useful to know what topics you will be learning about in the months to come, as well as the ones you are covering now. That way you can recognise useful opportunities when they arise.

The media are invaluable sources of information

Minimising problems

If you are fortunate, any problems you experience on your course will only be minor ones. For example, you may struggle to keep yourself motivated every single day and there may be times that you are having difficulty with a topic. Or you may be struggling to work with someone else in your team or to understand a particular tutor.

During induction you should have been told which tutor to talk to in this situation, and who to see if that person is absent or if you would prefer to see someone else. If you are having difficulties which are distracting you and affecting your work then it is sensible to ask to see your tutor promptly so that you can talk in confidence, rather than just trusting to luck everything will go right again. It is a rare student who is madly enthusiastic about every part of a course and all the other people on the course, so your tutor won't be surprised and will be able to give you useful guidance to help you stay on track.

If you are very unlucky, you may have a more serious personal problem to deal with. In this case it is important that you know the main sources of help in your school or college and how to access these.

- **Professional counselling** There may be a professional counselling service if you have a concern that you don't want to discuss with any teaching staff. If you book an appointment to see a counsellor then you can be certain that nothing you say will ever be mentioned to another member of staff without your permission.

- **Student complaint procedures** If you have a serious complaint to make then the first step is to talk to a tutor, but you should be aware of the formal student complaint procedures that exist if you cannot resolve the problem informally. Note that these are only used for serious issues, not for minor difficulties.

- **Student appeals procedures** If you cannot agree with a tutor about a final grade for an assignment then you need to check the grading criteria and ask the tutor to explain how the grade was awarded. If you are still unhappy then you should see your personal tutor. If you still disagree then you have the right to make a formal appeal.

- **Student disciplinary procedures** These exist so that all students who flout the rules in a school or college will be dealt with in the same way. Obviously it is wise to avoid getting into trouble at any time, but if you find yourself on the wrong side of the regulations do read the procedures carefully to see what could happen. Remember that being honest about what happened and making a swift apology is always the wisest course of action, rather than being devious or trying to blame someone else.

- **Serious illness** Whether this affects you or a close family member, it could severely affect your attendance. The sooner you discuss the problem with your tutor the better. This is because you will be missing notes and information from the first day you do not attend. Many students under-estimate the ability of their tutors to find inventive solutions in this type of situation – from sending notes by post to updating you electronically if you are well enough to cope with the work.

PLUSPOINTS	ACTION POINTS
+ Some students miss out on opportunities to learn more about relevant topics. This may be because they haven't read the unit specifications, so don't know what topics they will be learning about in future; haven't prepared in advance or don't take advantage of occasions when they can listen to an expert and perhaps ask questions. Examples of these occasions include external visits, visiting speakers, work experience, being at work and watching television.	✓ List the type of opportunities available on your course for obtaining more information and talking to experts. Then check with your tutor to make sure you haven't missed out any.
+ Many students encounter minor difficulties, especially if their course lasts a year or two. It is important to talk to your tutor, or another appropriate person, promptly if you have a worry that is affecting your work.	✓ Check out the content of each unit you will be studying so that you know the main topics you have still to study.
	✓ Identify the type of information you can find on television, in newspapers and in Podcasts that will be relevant to your studies.
+ All schools and colleges have procedures for dealing with important issues and problems such as serious complaints, major illnesses, student appeals and disciplinary matters. It is important to know what these are.	✓ Check out your school or college documents and procedures to make sure that you know who to talk to in a crisis and who you can see if the first person is absent.
	✓ Find out where you can read a copy of the main procedures in your school or college that might affect you if you have a serious problem. Then do so.

AND FINALLY . . .

Don't expect this Introduction to be of much use if you skim through it quickly and then put it to one side. Instead, refer to it whenever you need to remind yourself about something related to your course.

The same applies to the rest of this Student Guide. The Activities in the next section have been written to help you to demonstrate your understanding of many of the key topics contained in the core or specialist units you are studying. Your tutor may tell you to do these at certain times; otherwise there is nothing to stop you working through them yourself!

Similarly, the Marked Assignments in the final section have been written to show you how your assignments may be worded. You can also see the type of response that will achieve a Pass, Merit and Distinction – as well as the type of response that won't! Read these carefully and if any comment or grade puzzles you, ask your tutor to explain it.

Then keep this guide in a safe place so that you can use it whenever you need to refresh your memory. That way, you will get the very best out of your course – and yourself!

GLOSSARY

Note: all words highlighted in bold in the text are defined in the glossary.

Accreditation of Prior Learning (APL)

APL is an assessment process that enables your previous achievements and experiences to count towards your qualification providing your evidence is authentic, current, relevant and sufficient.

Apprenticeships

Schemes that enable you to work and earn money at the same time as you gain further qualifications (an **NVQ** award and a technical certificate) and improve your key skills. Apprentices learn work-based skills relevant to their job role and their chosen industry. You can find out more at www.apprenticeships.org.uk/

Assessment methods

Methods, such as **assignments**, case studies and practical tasks, used to check that your work demonstrates the learning and understanding required for your qualification.

Assessor

The tutor who marks or assesses your work.

Assignment

A complex task or mini-project set to meet specific **grading criteria**.

Awarding body

The organisation which is responsible for devising, assessing and issuing qualifications. The awarding body for all BTEC qualifications is Edexcel.

Core units

On a BTEC National course these are the compulsory or mandatory units that all students must complete to gain the qualification. Some BTEC qualifications have an over-arching title, eg Engineering, but within Engineering you can choose different routes. In this case you will study both common core units that are common to all engineering qualifications and **specialist core unit(s)** which are specific to your chosen **pathway**.

Degrees

These are higher education qualifications which are offered by universities and colleges. Foundation degrees take two years to complete; honours degrees may take three years or longer. See also **Higher National Certificates and Diplomas**.

DfES

The Department for Education and Skills: this is the government department responsible for education issues. You can find out more at www.dfes.gov.uk

Distance learning

This enables you to learn and/or study for a qualification without attending an Edexcel centre although you would normally be supported by a member of staff who works there. You communicate with your tutor and/or the centre that organises the distance learning programme by post, telephone or electronically.

Educational Maintenance Award (EMA)

This is a means-tested award which provides eligible students under 19, who are studying a full-time course at school or college, with a cash sum of money every week. See http://www.dfes.gov.uk/financialhelp/ema/ for up-to-date details.

External verification

Formal checking by a representative of Edexcel of the way a BTEC course is delivered. This includes sampling various assessments to check content and grading.

Final major project

This is a major, individual piece of work that is designed to enable you to demonstrate you have achieved several learning outcomes for a BTEC National qualification in the creative or performing arts. Like all assessments, this is internally assessed.

Forbidden combinations

Qualifications or units that cannot be taken simultaneously because their content is too similar.

GLH

See **Guided Learning Hours** on page 34.

Grade

The rating (Pass, Merit or Distinction) given to the mark you have obtained which identifies the standard you have achieved.

Grade boundaries

The pre-set points at which the total points you have earned for different units converts to the overall grade(s) for your qualification.

Grading criteria

The standard you have to demonstrate to obtain a particular grade in the unit, in other words, what you have to prove you can do.

Grading domains

The main areas of learning which support the **learning outcomes**. On a BTEC National course these are: application of knowledge and understanding; development of practical and technical skills; personal development for occupational roles; application of generic and **key skills**. Generic skills are basic skills needed wherever you work, such as the ability to work cooperatively as a member of a team.

Grading grid

The table in each unit of your BTEC qualification specification that sets out the **grading criteria**.

Guided Learning Hours (GLH)

The approximate time taken to deliver a unit which includes the time taken for direct teaching, instruction and assessment and for you to carry out directed assignments or directed individual study. It does not include any time you spend on private study or researching an assignment. The GLH determines the size of the unit. At BTEC National level, units are either 30, 60, 90 or 120 guided learning hours. By looking at the number of GLH a unit takes, you can see the size of the unit and how long it is likely to take you to learn and understand the topics it contains.

Higher education (HE)

Post-secondary and post-further education, usually provided by universities and colleges.

Higher level skills

Skills such as evaluating or critically assessing complex information that are more difficult than lower level skills such as writing a description or making out a list. You must be able to demonstrate higher level skills to achieve a Distinction grade.

Higher National Certificates and Diplomas

Higher National Certificates and Diplomas are vocational qualifications offered at colleges around the country. Certificates are part-time and designed to be studied by people who are already in work; students can use their work experiences to build on their learning. Diplomas are full-time courses – although often students will spend a whole year on work experience part way through their Diploma. Higher Nationals are roughly equivalent to half a degree.

Indicative reading

Recommended books and journals whose content is both suitable and relevant for the unit.

Induction

A short programme of events at the start of a course designed to give you essential information and introduce you to your fellow students and tutors so that you can settle down as quickly and easily as possible.

Internal verification

The quality checks carried out by nominated tutor(s) at your school or college to ensure that all assignments are at the right level and cover appropriate learning outcomes. The checks also ensure that all **assessors** are marking work consistently and to the same standard.

Investors in People (IIP)

A national quality standard which sets a level of good practice for the training and development of people. Organisations must demonstrate their commitment to achieve the standard.

Key skills

The transferable, essential skills you need both at work and to run your own life successfully. They are: literacy, numeracy, IT, problem solving, working with others and self-management.

Learning and Skills Council (LSC)

The government body responsible for planning and funding education and training for everyone aged over 16 in England – except university students. You can find out more at www.lsc.gov.uk

Learning outcomes

The knowledge and skills you must demonstrate to show that you have effectively learned a unit.

Learning support

Additional help that is available to all students in a school or college who have learning difficulties or other special needs. These include reasonable adjustments to help to reduce the effect of a disability or difficulty that would place a student at a substantial disadvantage in an assessment situation.

Levels of study

The depth, breadth and complexity of knowledge, understanding and skills required to achieve a qualification determines its level. Level 2 is broadly equivalent to GCSE level (grades A*-C) and level 3 equates to GCE level. As you successfully achieve one level, you can then progress on to the next. BTEC qualifications are offered at Entry level, then levels 1, 2, 3, 4 and 5.

Local Education Authority (LEA)

The local government body responsible for providing education for students of compulsory school age in your area.

Mentor

A more experienced person who will guide and counsel you if you have a problem or difficulty.

Mode of delivery

The way in which a qualification is offered to students, eg part-time, full-time, as a short course or by **distance learning**.

National Occupational Standard (NOS)

These are statements of the skills, knowledge and understanding you need to develop to be competent at a particular job. These are drawn up by the **Sector Skills Councils**.

National Qualification Framework (NQF)

The framework into which all accredited qualifications in the UK are placed. Each is awarded a level based on their difficulty which ensures that all those at the same level are of the same standard. (See also **levels of study**.)

National Vocational Qualification (NVQ)

Qualifications which concentrate upon the practical skills and knowledge required to do a job competently. They are usually assessed in the workplace and range from level 1 (the lowest) to level 5 (the highest).

Nested qualifications

Qualifications which have 'common' units, so that students can easily progress from one to another by adding on more units, such as the BTEC Award, BTEC Certificate and BTEC Diploma.

Pathway

All BTEC National qualifications are comprised of a small number of core units and a larger number of specialist units. These specialist units are grouped into different combinations to provide alternative pathways to achieving the qualification, linked to different career preferences.

Peer review

An occasion when you give feedback on the performance of other members in your team and they, in turn, comment on your performance.

Plagiarism

The practice of copying someone else's work and passing it off as your own. *This is strictly forbidden on all courses.*

Portfolio

A collection of work compiled by a student, usually as evidence of learning to produce for an **assessor**.

Professional body

An organisation that exists to promote or support a particular profession, such as the Law Society and the Royal Institute of British Architects.

Professional development and training

Activities that you can undertake, relevant to your job, that will increase and/or update your knowledge and skills.

Project

A comprehensive piece of work which normally involves original research and investigation either by an individual or a team. The findings and results may be presented in writing and summarised in a presentation.

Qualifications and Curriculum Authority (QCA)

The public body, sponsored by the **DfES**, responsible for maintaining and developing the national curriculum and associated assessments, tests and examinations. It also accredits and monitors qualifications in colleges and at work. You can find out more at www.qca.gov.uk

Quality assurance

In education, this is the process of continually checking that a course of study is meeting the specific requirements set down by the awarding body.

Sector Skills Councils (SSCs)

The 25 employer-led, independent organisations that are responsible for improving workforce skills in the UK by identifying skill gaps and improving learning in the workplace. Each council covers a different type of industry and develops its **National Occupational Standards**.

Semester

Many universities and colleges divide their academic year into two halves or semesters, one from September to January and one from February to July.

Seminar

A learning event between a group of students and a tutor. This may be student-led, following research into a topic which has been introduced earlier.

Specialist core units

See under **Core units**.

Study buddy

A person in your group or class who takes notes for you and keeps you informed of important developments if you are absent. You do the same in return.

Time-constrained assignment

An assessment you must complete within a fixed time limit.

Tutorial

An individual or small group meeting with your tutor at which you can discuss the work you are currently doing and other more general course issues. At an individual tutorial your progress on the course will be discussed and you can also raise any concerns or personal worries you have.

The University and Colleges Admissions Service (UCAS)

The central organisation which processes all applications for higher education courses. You pronounce this 'You-Cass'.

UCAS points

The number of points allocated by **UCAS** for the qualifications you have obtained. **HE** institutions specify how many points you need to be accepted on the courses they offer. You can find out more at www.ucas.com

Unit abstract

The summary at the start of each BTEC unit that tells you what the unit is about.

Unit content

Details about the topics covered by the unit and the knowledge and skills you need to complete it.

Unit points

The number of points you have gained when you complete a unit. These depend upon the grade you achieve (Pass, Merit or Distinction) and the size of the unit as determined by its **guided learning hours**.

Vocational qualification

A qualification which is designed to develop the specific knowledge and understanding relevant to a chosen area of work.

Work experience

Any time you spend on an employer's premises when you carry out work-based tasks as an employee but also learn about the enterprise and develop your skills and knowledge.

ACTIVITIES

This section focuses on grading criteria P1, P2, P3, P4, P5, P6, P7; M1, M2, M3 and aspects of D1 and D2 from Unit 2 – Communication for Technicians.

Learning outcomes

1 Be able to interpret and use simple engineering drawings/circuit/network diagrams and sketches to communicate technical information

2 Be able to use verbal and written communication skills in engineering settings

3 Be able to obtain and use engineering information

4 Be able to use information and communication technology (ICT) to present information in engineering settings

Content

1) Be able to interpret and use simple engineering drawings/circuit/network diagrams and sketches to communicate technical information

Interpret: obtain information and describe features, eg component features, dimensions and tolerances, surface finish; identify manufacturing/assembly/ process instructions, eg cutting lists, assembly arrangements, plant/process layout or operating procedures, electrical/ electronic/communication circuit requirements; graphical information used to aid understanding of written or verbal communication, eg illustrations, technical diagrams, sketches.

Engineering drawings/circuit/network diagrams: working documents, eg first and third angle projections, detail and assembly drawings, plant/process layout diagrams, electrical/electronic/communications/ circuit diagrams, system/network diagrams; use of common drawing/circuit/network diagram conventions and standards, eg layout and presentation, line types, hatching, dimensions and tolerances, surface finish, symbols, parts lists, circuit/component symbols, use of appropriate standards (British (BSI), International (ISO)).

Sketches: free-hand illustration of engineering arrangements using 2D and 3D techniques, eg components, engineering plant or equipment layout, electrical/ electronic circuits/network diagrams, designs or installations.

2) Be able to use verbal and written communication skills in engineering settings

Written work: note taking, eg lists, mind mapping/flow diagrams; writing style, eg business letter, memo writing, report styles and format, email, fax; proofreading and amending text; use of diary/logbook for planning and prioritising work schedules; graphical presentation techniques, eg use of graphs, charts and diagrams.

Verbal methods: speaking, eg with peers, supervisors, use of appropriate technical language, tone and manner; listening, eg use of paraphrasing and note taking to clarify meaning; impact and use of body language in verbal communication.

3) Be able to obtain and use engineering information

Information sources: non-computer-based sources, eg books, technical reports, institute and trade journals, data sheets and test/experimental results data, manufacturers' catalogues; computer-based sources, eg inter/intranet, CD ROM-based information (manuals, data, analytical software, manufacturers' catalogues), spreadsheets, databases.

Use of information: eg for the solution of engineering problems, for product/service/ topic research, gathering data or material to support own work, checking validity of own work/findings.

4) Be able to use information and communication technology (ICT) to present information in engineering settings

Software packages: word processing; drawing, eg 2D CAD, graphics package; data handling and processing, eg database, spreadsheet, presentation package, simulation package such as electrical/ electronic circuits, plant/process systems;

communication, eg email, fax, inter/intranet, video conferencing, optical and speech recognition system.

Hardware devices: computer system, eg personal computer, network, plant/process control system; input/output devices, eg keyboard, scanner, optical/speech recognition device, printer, plotter.

Present information: report that includes written and technical data, eg letters, memos, technical product/service specification, fax/email, tabulated test data, graphical data; visual presentation, eg overhead transparencies, charts, computer-based presentations (PowerPoint).

Grading criteria

P1 interpret an engineering drawing/circuit/ network diagram and sketches

Engineers use several different types of drawing in order to communicate their ideas. These include formal drawings as well as less formal diagrams and sketches. You will need to be able to read and interpret a variety of different types of engineering drawings to obtain information or to understand a task. You will also need to be able to interpret sketches and other types of illustration, such as exploded views.

P2 produce an engineering drawing/circuit/ network diagram and sketches

You will need to demonstrate that you are able to produce a variety of different types of engineering drawing. These include formal drawings as well as schematic diagrams used in electrical engineering, electronics and in pneumatic and hydraulic applications. You should also be able to produce freehand 2D and 3D sketches of engineering components. These sketches should be of sufficient quality to communicate essential information about a component such as outline shape, form and dimensions. You should also be able to produce block schematics, flowcharts, circuits and layouts using the appropriate hand drawing and sketching techniques.

P3 identify and use appropriate standards, symbols and conventions in the production of an engineering drawing/circuit/network diagram

The drawings and technical illustrations that you produce need to use standard symbols and drawing conventions. You will need to be familiar with relevant British Standards for the production of engineering drawings.

P4 communicate information effectively in written work

You will need to be able to demonstrate that you can communicate with engineers in writing. You should be able to prepare a written technical instruction (for example, how to assemble a component or how to carry out a test or measurement) as well as a technical report. You also need to be able to show that you can use appropriate language, spelling and grammar in written communications, including letters and email messages.

P5 communicate information effectively using verbal methods

As well as using written and graphical communication, engineers frequently need to talk to one another and clients: verbal communication. This might involve giving and receiving instructions, as well as dealing with a telephone message, talking to clients and customers and giving presentations. Engineers also need to be involved with face-to-face meetings and interviews. All of these mean that you should be as proficient in verbal communication as in your writing, drawing and sketching.

P6 identify and use appropriate information sources to solve an engineering task

You need to be able to use a wide variety of information sources used by engineers (technical books, datasheets, literature, CD ROM and on-line information sources) to obtain data relating to an engineered product. You also need to be able to summarise the information that you have gathered (from both electronic and non-electronic sources) in the form of a brief technical report with appropriate recommendations. Note that, whenever you do this, it is important to quote your original reference sources.

P7 select and use appropriate ICT software packages and hardware devices to present information

You need to be able to use the right hardware and software to present information using information and communication technology (ICT). This means selecting a suitable software package for the task that you want to perform. For example, a basic 2D CAD package would be perfectly adequate for producing a simple isometric drawing but would normally be incapable of producing a fully rendered 3D view of a component.

Similar considerations apply to hardware. For example, a hand-held scanner would be appropriate for scanning text from a technical report for optical character recognition but would be inadequate for the accurate scanning of an A4 drawing sheet.

M1 evaluate a written communication method and identify ways in which it could be improved

You need to be able to think about different methods of written communication, compare them and identify ways in which they can be improved. For example, you might find that several paragraphs of text could be replaced by a series of bullet points, which are much easier to read. Equally, you might decide that a basic list of steps is insufficient to explain a series of complex processes and that a better way of presenting this information might be to use a series of annotated diagrams.

M2 review the information sources obtained to solve an engineering task and explain why some sources have been used but others rejected

Some information sources may be more reliable or more detailed than others, and some may present information in a more complex manner than is actually required. It is therefore important to be able to exercise some critical judgement as to which source (or sources) should be used to solve a particular problem or to gain a particular piece of information.

You need to be able to justify your choice of a specific source of information (including both electronic and non-electronic sources). This means giving reasons for using the source that you have chosen as well as the reasons for not selecting alternative sources. For example, you might find it best to use an on-line datasheet rather than a printed data book. Your reasons for choosing the on-line datasheet might be based on the fact that the product specification is being regularly updated and that a printed

data book may no longer contain current information.

M3 use an ICT software package and its tools to prepare and present clearly laid out work

You will need to use ICT software (such as AutoSketch, AutoCAD or TurboCAD) to produce your formal engineering drawings. If you need to produce technical illustrations (rather than formal 2D and 3D engineering drawings) alternative software can often be quicker and easier to use than a conventional CAD package. Your school or college may be able to provide you with access to packages such as MS Visio, Corel Draw and SmartDraw. All of these packages support the use of templates and symbol libraries and are well worth investigating.

For the production of presentations, most schools and colleges will be able to provide you with access to MS PowerPoint. This package will allow you to produce good-looking presentations that can incorporate text, graphics, animation and video clips. And don't forget that if you have a digital camera, this can be an excellent tool for augmenting your presentations with some original high-quality photographic material!

D1 justify their choice of a specific communication method and the reasons for not using a possible alternative

Finally, you need to be able to justify the choice of a specific method of communication. Why is the method that you have chosen the best one for the task in hand? This means giving reasons for using the method that you have chosen as well as the reasons for not selecting an alternative method. For example, you might find it best to use a PowerPoint presentation rather than a written report. Your reasons for choosing PowerPoint might focus on the need to incorporate some animated material plus a short video sequence. This would be impossible in a printed document and prohibitively expensive if it had to be produced as a conventional video or DVD.

Other tasks that you may need to be involved with include:

- delivering a brief presentation (of eight minutes or more) using appropriate visual aids and responding appropriately to questions

- conducting a brief interview (lasting no longer than 15 minutes) with another student and taking notes to summarise the outcome

- taking part in a group discussion to identify or share technical information within a set task

- preparing a letter to an engineering supplier requesting modifications to an engineered component

- preparing a brief technical report concerning a design modification

- producing a data sheet for a simple engineered product or service

- using information sources (literature, CD ROM and on-line) to obtain data relating to an engineered product and summarising this in the form of a brief technical report

- sending and receiving email correspondence to convey engineering ideas and technical data.

D2 critically evaluate their use of an ICT presentation method and identify an alternative approach

In connection with your ICT presentation, you will need to consider whether or not you have used an appropriate method of communicating information. For example, would it have been better to use a paper-based, word-processed report or perhaps an electronic document produced in Adobe Acrobat portable document format (PDF)? You might also consider whether the information could be presented as a series of web pages accessible using a web browser. When doing this you might want to use some form of table or chart that lists the advantages and disadvantages of each method.

ACTIVITY 1

A detailed engineering drawing of a miniature electric motor is shown in figure 1. Use the drawing to answer the following questions:

1 What are the approximate overall dimensions of the motor?

2 What is the name of the company that manufactures the motor?

3 What is the drawing reference and drawing number?

4 What units are used to indicate the dimensions of the motor?

5. What projection has been used to produce the drawing?

6 What is the distance between the two electrical terminal connections at the base of the motor?

7 What is the diameter of the motor's output shaft?

All of the information that you need to answer questions 1 to 7 can be obtained from the drawing. Engineers use drawings like this on a daily basis and you need to be very familiar with the techniques and conventions used in their production.

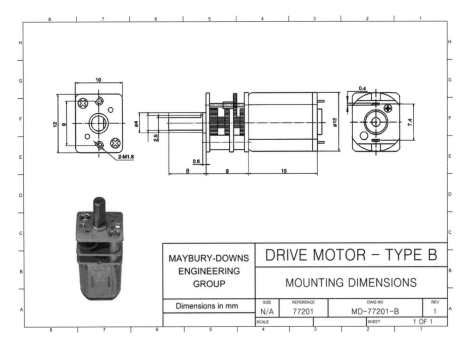

ACTIVITY 2

Figure 1 shows the circuit diagram of a 5 V power supply. The printed circuit board layout of the power supply is shown in figure 2. Use these diagrams to answer the following questions:

1   How many connections are there on SK1 and SK2?
2   What integrated circuit device is used for U1 and how many pins does it have?
3   Which five components are connected to the negative output (pins 2 and 3 of SK2)?
4   What component is connected in parallel with C1?
5   How many individual diodes are there inside BR1?
6   What type of component is C2 and what is its value?
7   Which pin is unused on SK1?
8   Which is the largest component?

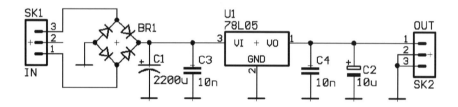

Figure 1 Circuit diagram for the 5 V power supply

44

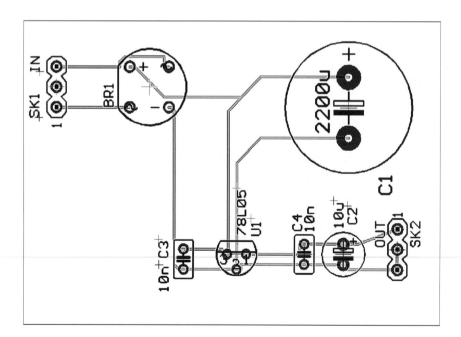

Figure 2 Printed circuit board component layout for the 5 V power supply

ACTIVITY 3

The photograph shown in figure 1 shows the nose undercarriage leg of Concorde, the world's first supersonic passenger aircraft. Make freehand pencil sketches of the undercarriage viewed (a) from the front of the aircraft (right of the photograph) and (b) from the starboard side (left of the photograph).

Figure 1 Nose undercarriage leg of the Concorde supersonic passenger aircraft

You need to start this exercise with a clean sheet of A4 drawing paper and an H-pencil. To help you get the proportions right it can be useful to imagine a rectangular grid superimposed over the picture, as shown in figure 2. Start by sketching the main undercarriage strut (we've aligned this with the grid in figure 2) and then add the bracing struts, wheels and other components. Don't worry if your first attempt doesn't look too good, you might need to make several sketches before you are happy with the final result.

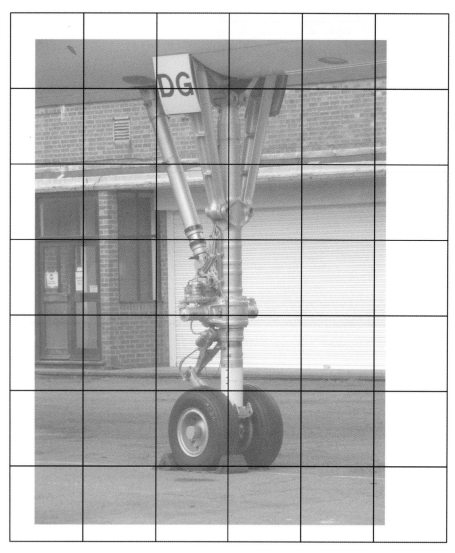

Figure 2 Superimposing an imaginary grid over figure 1 will help you get the proportions right when sketching

46

ACTIVITY 4

Use an isometric grid to sketch isometric views of the three components shown in figure 1. (Don't worry about the dimensions; draw them to the scale that suits you best.)

This task is easily achieved using an isometric grid. An example of how the component shown in figure 1(a) should be drawn is shown in figure 2.

(a)

(b)

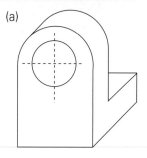

(c)

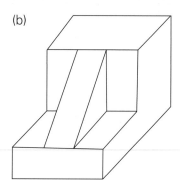

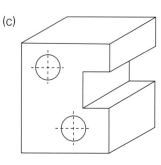

Figure 1 Three components shown in oblique projection

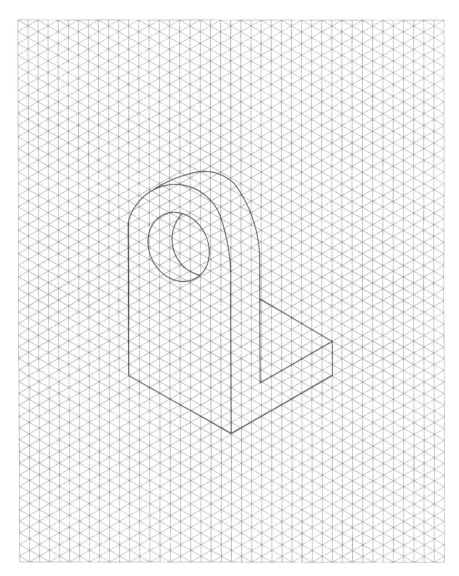

Figure 2 The completed isometric view for the component shown in figure 1(a)

ACTIVITY 5

Obtain relevant British Standards for drawing (your school or college should have copies of these) and use these to answer the following questions:

1 Read the section on dimensioning and identify the **four** important principles of dimensioning.

2 Sketch an example of a functional dimension and briefly explain its purpose.

3 Sketch the necessary projection lines and show an example of how to dimension points of intersection.

4 Sketch an example of an auxiliary dimension and briefly explain its purpose.

5 Sketch an example of dimensioning by co-ordinates and suggest what the main purpose of this is.

47

6 Sketch an example of dimensioning for each of the following features:

(a) an internal radius

(b) an external radius

(c) a countersink

(d) a counterbore

(e) an external chamfer

(f) a spotface

(g) an internal screw thread

(h) an external screw thread

(i) six holes equally spaced on a pitch circle diameter of 60 mm

(j) an external taper.

Present your work in the form of a brief report entitled 'Dimensioning Engineering Drawings'.

ACTIVITY 6

Produce a laminated A4 instruction card on any one of the following topics:

- safety information when using a soldering iron
- safety information when connecting a car battery charger
- adjusting and using a micrometer
- adjusting and using a torque wrench
- adjusting and using a multimeter
- adjusting and using an oscilloscope.

The card is to be displayed in a workshop and should be designed for use by students with no previous experience. Include any relevant safety information.

You need to think about this activity before you get started. You only have an A4 page available so this will limit the amount of text that can be used. You will also need to ensure that the text is readable. This means that you will need to consider what typeface and font size you will use. Also, because you want people to read your card, it will need to be made 'eye-catching'. This means using attractive colours and including some relevant illustrations or clip-art. You could start this activity by doing some rough design sketches and asking your friends and/or your tutor to comment on them. You might also want to look at some existing instruction cards to see what works and what doesn't!

Next you need to make a list of some of the bullet points that you will need to include in your instruction card. For example, in the case of safety information when using a soldering iron you might want to include some or all of the following points:

Safety information when using a soldering iron

- Always ensure that the soldering iron is placed in its holder when not in use.

- Check that the soldering iron is used on a soldering mat or other heat-proof surface.

- Do not leave a soldering iron switched on for long periods when it is not being used.

- Use low-voltage soldering equipment whenever possible.

- Always check the condition of the soldering iron, holder and lead before use.

- Check that the soldering iron bit is correct for the work to be carried out.

- Always check that the correct soldering temperature has been set.

- Allow the soldering iron to heat up and reach its normal working temperature before use.

Take care to avoid these hazards:

- Solder fumes are an irritant and exposure, particularly if prolonged, can cause asthma attacks. Always ensure that the area is well ventilated and use fume extraction equipment when available.

- Molten solder or flux additives can cause permanent eye damage. Always use eye protection (safety glasses or a bench magnifier for close work).

- When working on sensitive circuits, check that the soldering iron is protected against ESD (electrostatic discharge).

- An electric shock hazard may exist if the lead to a mains-operated soldering iron or its mains connector becomes loose or frayed.

- Always ensure that electronic equipment is switched off and disconnected from the mains supply (or batteries removed) before attempting to use a soldering iron to remove or replace wiring or components.

- The bit of a soldering iron is usually maintained at a temperature of between 250°C and 350°C. At this temperature, conventional plastic insulating materials will melt and many other materials (such as paper, cotton, etc) will burn. Personal contact with the bit and heated parts should be avoided at all times.

ACTIVITY 7

Use a standard engineer's Vee-block to produce a short verbal presentation (no more than 5 minutes) with relevant handouts to explain the difference between isometric and oblique projection.

A Vee-block makes an excellent visual aid for showing the difference between these two projections that are commonly used in engineering drawing. The Vee-block is small, easily recognisable and can be turned around to show its different faces and views that are possible when the tool is drawn in the two projections.

Your presentation could be based on the following points and flip-chart diagrams or overhead projector transparencies:

Isometric projection

- Shows how an object appears when viewed from different angles
- Uses two receder lines at 30° to the horizontal plus a vertical line
- A disadvantage of the technique is that circles and arcs can be difficult to draw
- An advantage of the technique is that components look realistic
- Drawings are simplified by using a standard isometric grid which is composed of lines drawn at 30°. The grid forms a series of equilateral triangles (usually with a side of length 5 mm).

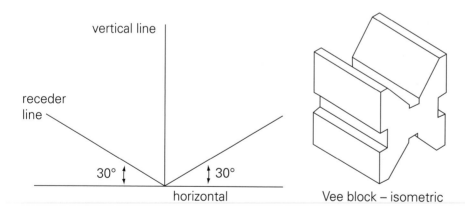

Figure 1 Isometric projection

Oblique projection

- Shows how an object appears when viewed from different angles
- Uses one receder line at 45° plus vertical and horizontal axes
- An advantage of this technique is that the front face of the object appears 'flat' (useful if this face has circles or arcs)
- A disadvantage of this technique is that drawings can look 'out of proportion' if lines along the receder are drawn full length (they are often shortened to half or two-thirds size in order to overcome this problem).

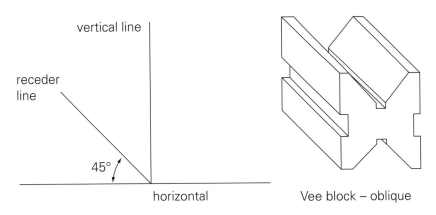

Figure 2 Oblique projection

ACTIVITY 8

You are completing a short work placement in a company that carries out repairs and modifications to commercial vehicles. The workshop supervisor, Dan Evans, needs to produce a safety notice with guidance for personnel using welding equipment. He has produced a written draft and he has asked you to improve it. Here is Dan's draft text:

Important workshop safety notice

Please read this before carrying out any gas welding in this workshop

The use of oxy/fuel/gas equipment should not be used unless it has been authorised by the workshop supervisor or another authorised person. You also need to be trained to use the equipment. Nobody should use it unless they have had training in the safe use of the equipment and taking precautions to avoid safety hazards. Training in fire extinguisher use is also required and the means of escape, raising the fire alarm and calling the fire brigade when a fire breaks out.

51

Before use check that the equipment is in good condition. It consists of cylinders of oxygen and fuel gas (propane or acetylene) and a means shutting off or isolating the gas supply (usually the cylinder valves) plus a pressure regulator fitted to the outlet valve of the gas cylinder (used to reduce and control gas pressure) and a flashback arrester to protect cylinders from flashbacks and backfires. There is also a flexible hose that conveys the gas from the cylinder to the blowpipe and a on-return valve that stops oxygen reverse flow back into the oxygen line. Finally you will see the blowpipe where the gas is mixed with the oxygen and ignited. Check each part for condition and serviceability before use.

Notes:

Hose colours: Acetylene hose is red and oxygen hose is black.

Cylinder colours: Acetylene cylinder is maroon and oxygen cylinder is black.

Safety when welding

A lighted blowpipe is a very dangerous piece of equipment. To avoid injury you should work well away from other people and you should wear protective clothing and goggles for eye protection. You should also shut off the blowpipe when not in use. Do not leave a lighted blowpipe on a bench or the floor as the force of the flame may cause it to move. Also you need to clamp the workpiece not hold it in your hands. Keeping hoses away from the working area to prevent accidental contact with flames, heat, sparks or hot spatter is also important.

Dan Evans

Workshop Supervisor

You should read Dan's draft text carefully and decide on which points need to be made prominent and how the text can be improved to make it more readable and succinct. You might want to consider rearranging the text and using bullet points in places rather than sentences. You might also want to consider ways of emphasising safety-critical points.

Write short notes to indicate where the problems are and what should be done to put things right (you will need to feed this information back to Dan when you show him your completed version).

Finally, you should produce your own improved version of the workshop safety notice. This should be word-processed and printed on a single A4 sheet, suitable for enlargement using a photocopier to make an A3 poster for display in a prominent place in the workshop. Make sure that you use appropriate font styles and colours and ensure that the language can be understood by anyone that might need to read the information.

ACTIVITY 9

You are on a work placement at Howard Associates, a company that supplies electronic equipment and components. Your supervisor, John Jones, has asked you to deal with the email enquiry from a customer shown in figure 1. Use the Howard Associates datasheet for the 74ALS08 (shown in figures 2 and 3) to prepare an email reply for Darren Baker that answers each of his questions.

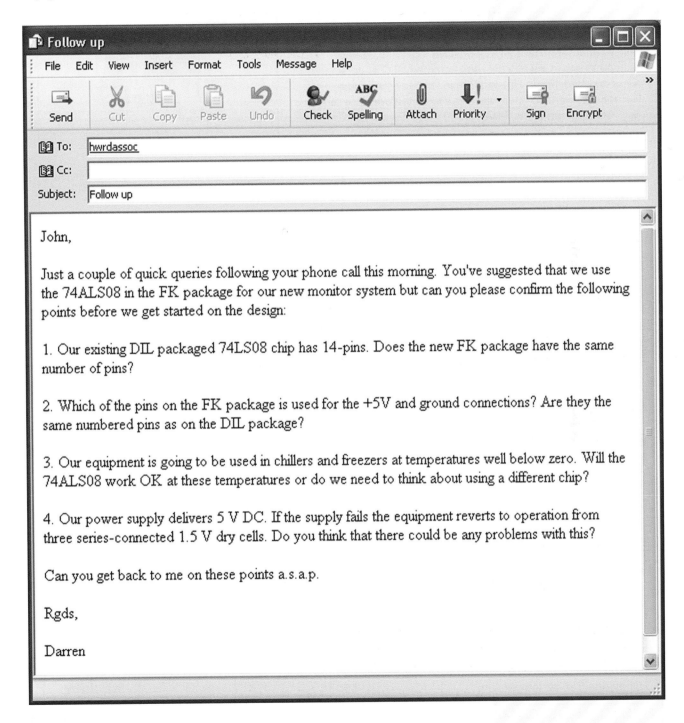

Figure 1 Darren Baker's email to John Jones at Howard Associates

Howard Associates

DATA SHEET	54ALS08, 54AS08, 74ALS08, 74AS08
	Quadruple 2-input AND gates

MAIN FEATURES

- Four independent 2-input positive logic AND gates
- Available in military (54) and commercial (74) versions
- Standard 14-pin DIL and 20-pin leadless packages
- Military version operates over a wide temperature range
- Complementary to 54/74ALS00 quad 2-input NAND gate

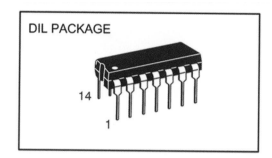

DIL PACKAGE

14

1

TRUTH TABLE
(each gate)

INPUTS		OUTPUT
A	**B**	**Y**
H	H	H
L	X	L
X	L	L

H = logic 1 (high)
L = logic 0 (low)
X = don't care (either high or low)

SYMBOL

1A 1
1B 2
3 1Y

2A 4
2B 5
6 2Y

3A 9
3B 10
8 3Y

4A 12
4B 13
11 4Y

ELECTRICAL CHARACTERISTICS (T_A = +25°C)

Parameter	Symbol	54ALS08			74ALS08			Unit
		Min.	Nom.	Max.	Min.	Nom.	Max.	
Supply voltage	V_{CC}	4.5	5.0	5.5	4.5	5.0	5.5	V
High-level input voltage	V_{IH}	2.0			2.0			V
Low-level input voltage	V_{IL}			0.8			0.8	V
High-level output current	I_{OH}			−0.4			−0.4	mA
Low-level output current	I_{OL}			4.0			8.0	mA
Operating free-air temperature	T_A	−55		125	0		70	°C

Data sheet reference: 95-072
page 1

Figure 2 Page 1 of the Howard Associates datasheet for the 74ALS08

ABSOLUTE MAXIMUM RATINGS (T_A = +25°C)

Supply voltage, V_{CC}		7 V
Input voltage, V_I		7 V
Operating free-air temperature range, T_A:	54ALS08	−55°C to +125°C
	74ALS08	0°C to +70°C
Storage temperature, T_{stg}		−65°C to +150°C

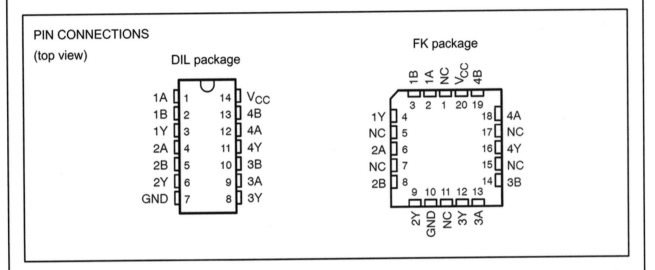

PIN CONNECTIONS
(top view)

ELECTRICAL CHARACTERISTICS

Parameter	Test conditions		54ALS08			74ALS08			Unit
			Min.	†Typ.	Max.	Min.	†Typ.	Max.	
V_{IK}	V_{CC} = 4.5 V	I_I = −18 mA			−1.5			−1.5	V
V_{IH}	V_{CC} = 5 V	I_{OH} = −0.4 mA	3.0			3.0			V
V_{OL}	V_{CC} = 4.5 V	I_{OL} = 4 mA		0.25	0.4		0.25	0.4	V
		I_{OL} = 8 mA					0.35	0.5	V
I_I	V_{CC} = 5.5 V	V_I = 7 V			0.1			0.1	mA
I_{IH}	V_{CC} = 5.5 V	V_I = 2.7 V			20			20	µA
I_{IL}	V_{CC} = 5.5 V	V_I = 0.4 V			−0.1			−0.1	mA
I_O	V_{CC} = 5.5 V	V_O = 2.25 V	−20		−112	−30		−112	mA
I_{CCH}	V_{CC} = 5.5 V	V_I = 4.5 V		1.3	2.4		1.3	2.4	mA
I_{CCL}	V_{CC} = 5.5 V	V_I = 0 V		2.2	4.0		2.2	4.0	mA

†Typical values are at V_{CC} = 5 V, T_A = +25°C

Data sheet reference: 95-072
page 2

© Howard Associates 2007

Figure 2 Page 2 of the Howard Associates datasheet for the 74ALS08

ACTIVITY 10

Use the MatSdata materials database for students (available for free download from www.key2study.com) to locate information on:

- ABS (fibre-reinforced thermoplastic)
- carbon fibre composite.

Use the data to compare and contrast these two materials in terms of their suitability for use as the hull of a small power boat. You should consider properties such as density, strength, cost factor, processes and techniques in your answer.

What is the composition of each of these materials?

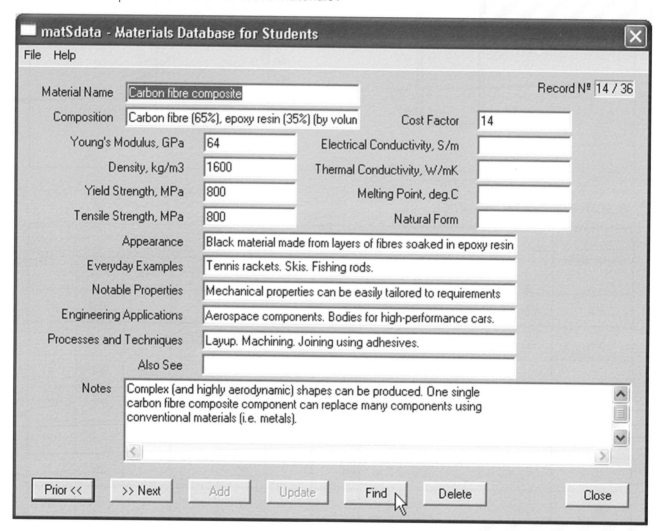

Figure 1 The MatSdata entry for carbon fibre composite material

ACTIVITY 11

Howard Aerospace has asked you to advise them on the potential for replacing the conventional tail lights (which use filament lamps) on a light aircraft with a unit that uses high-intensity light emitting diodes (LED). They have obtained literature from two suppliers: Airlight Aviation Components (AAC) and Napsos Engineering Ltd (NEL). Howard Aerospace has asked you to comment on this and suggest (with reasons) which company's product should be further investigated.

The outline specification sheet supplied by AAC is as follows:

AAC BCX-190 TAIL LIGHT OPERATING AND INSTALLATION INSTRUCTIONS

Nominal Voltage: 28 VDC
Operational Voltages: 22 – 30 VDC
Input Current: 0.30 Amps

EQUIPMENT LIMITATIONS: The approved tail position light assembly should be mounted vertically and as far aft on the aircraft as possible, on a thermally conductive surface.

CONTINUED AIRWORTHINESS: The LED tail position light assembly is designed with six individual LEDs. If any one LED fails, the unit must be repaired or replaced.

INSTALLATION PROCEDURES:

1 Choose the appropriate model light assembly which is most applicable to your aircraft.
2 Remove the old light, locate and save the existing +28VDC lead and (−) ground lead. Clean and prepare ends as required.
3 Use the existing mounting holes and hardware. Note: If a special adaptor plate is required, contact us with your requirements.
4 Connect the existing +28 VDC lead to the white 20 AWG lead on the input cable assembly (supplied with light assembly). Connect the existing ground lead to the black 20 AWG wire on the input cable assembly. If a chassis ground wire exists on the aircraft, connect it to the chassis ground wire assembly. Both leads must be connected by JAA/FAA/EASA approved techniques using approved hardware.
5 Make sure the existing system is equipped with an appropriate sized breaker.
6 Remove two #4–40 screws from tail light assembly. Save these screws.
7 Install the light assembly and ensure that all the leads are clear of any obstructions and tie-wrap as required. Secure light assembly using approved vibration resistant hardware methods. Note "Top" marking on the unit.
8 Re-mount the lens onto the assembly. Attach lens with mounting screws (provided). NOTE: These screws feature a nylon thread locking patch.
9 Check all avionics systems for interference from this installation.
10 When waterproofing the assembly to the aircraft, apply aviation approved single part silicone (RTV) or equivalent around the periphery with caution. Do not fill the drain holes at bottom of unit. Note: RTV should never be used between the rear surface of the light assembly and its mounting surface.
11 Update aircraft records, complete Form 337 and obtain regulatory body field approval for installation.

The NEL specification is as follows:

LVF20

Low-current, high-efficiency, high-reliability tail light unit

The NEL LVF20 tail light unit comprises an array of 10 high-intensity, high-efficiency LEDs. The unit requires a standard aircraft DC supply, is easy to fit and is designed to replace the existing tail light on most single and twin engine light aircraft.

To fit an NEL light unit:

(a) Remove the existing light unit and ensure that the NEL unit can be fitted in its place. Various fitting configurations are possible – see manual for more information.

(b) Connect the +28V supply cable to your aircraft's positive DC supply (ensure that the aircraft supply is isolated before you do this).

(c) Connect the negative lead to the aircraft's bonded ground system.

(d) Remove the protective film from the lens cover.

(e) Restore the supply to the aircraft tail light and perform a ground check.

NB: You must ensure that you comply with airworthiness requirements when fitting an NEL tail light.

For further information contact Paul Smith on 0779 223581

Read the two documents carefully and compare the information contained in them. Comment on the quality and sufficiency of information provided by the two companies and decide on which of the two company's products should be further investigated and which should now be rejected. Write a brief note to Howard Associates, clearly stating the reasons for your decision.

ACTIVITY 12

Figure 1 shows four different types of engineering component. Draw each component using a 2D CAD package and a standard drawing template. With reference to British Standard practice, add the necessary dimensions to ensure that the components could be manufactured from your drawing (you can ignore the thickness of components (b), (c) and (d) for the purposes of this exercise).

(a) Turned shaft

(b) Plate

(c) Gland

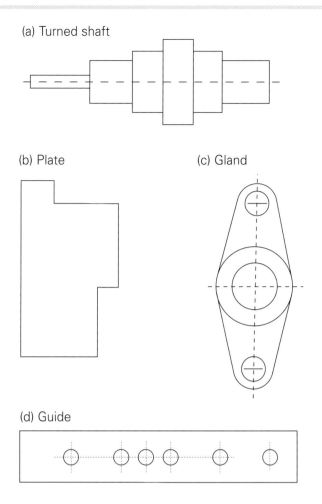

(d) Guide

Figure 1 Four engineering components

Before you start this exercise it is important to check that you understand how to indicate dimensions using recommended British Standard practice.

ACTIVITY 13

Figure 1 shows a cast steel component. Use a 2D CAD package to draw this component using orthographic projection (either first or third angle) using a standard drawing sheet template.

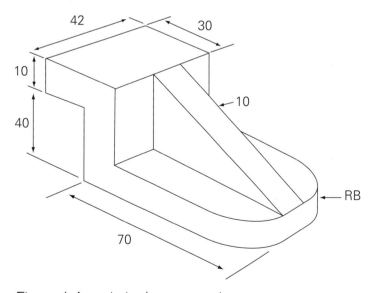

Figure 1 A cast steel component

You should have spotted that the sketch uses isometric projection to provide a 3D view of the component. Orthographic projection requires you to produce three separate views (often referred to as elevation, plan and end views). You can do this by viewing the component from the left side and then drawing (flat) the view that you would see. Next you need to create the plan view (it is usually convenient to place this underneath the elevation that you have just produced). The plan view is the view that you would see (flat) when positioned above the component. Finally, the end view can be drawn. This is the view from the far side, looking back at the component.

Since this is a formal engineering drawing, it's necessary to use a standard drawing sheet template. You will probably have already created such a template as a previous drawing exercise. Figure 2 shows how the final drawing might look.

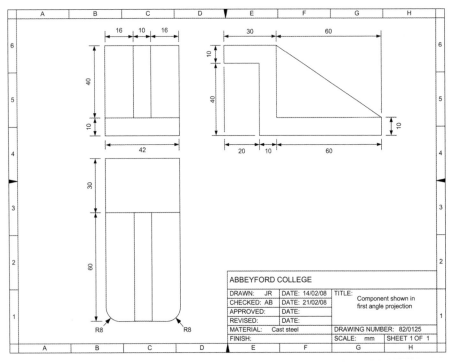

Figure 2 Completed orthographic projection of the cast steel component

ACTIVITY 14

Task 1

Prepare and deliver a 15-minute (including time for questions) presentation on any one of the following topics:
- using a digital camera
- 3G mobile phones
- preparing your car for an MOT test
- wireless networks
- fitting a sound card to a computer
- using a soldering iron
- choosing an MP3 player.

Make sure that there is a good selection of topics across your class group.

You should decide what format your presentation will take. Flip chart? Microsoft PowerPoint? Some other format?

Your presentation should be structured (it should have a beginning, middle and end): it should have an introduction, with the information that you want to present, and a summary of your findings or recommendations. As part of your introduction it is a good idea to explain why you chose the topic in the first place and why it is important for your audience to know about it.

Then write the main content of your presentation and some sort of summary to finish off.

Don't forget to introduce yourself and don't forget to thank your audience when you've finished and invite any questions from them.

A PowerPoint presentation should include bullet points, brief text, relevant diagrams, charts and graphs. For a 10-minute presentation you should use about five or six slides. Any more than that and you will be flicking between the slides far too quickly for your audience. The standard templates provided as part of the PowerPoint package will provide you with plenty of choice concerning style and colour.

You should aim to maintain the attention of your audience by using an appropriate pace for your delivery (not so fast that your audience has difficulty following what you are saying to them and showing them and not so slow that your audience begins to get bored). You should carefully run through your presentation several times before you deliver it and you might want to elicit some help and advice from a 'tame' audience.

You should make it clear at the start of the presentation whether you will be inviting questions at the end or dealing with them as you go along.

Finally, if you are worried about having to make a presentation 'in public' remember that this is something that most of us have to do at some time in our lives. So, try to use the activity as something that you will learn from. The next presentation that you have to give will be just that little bit easier!

Task 2

How do you think the use of technology in your presentation helped you get your meaning across?

Think about the same question for presentations by others in your class.

Discuss how you might get your meaning across using a different presentation method.

This section focuses on grading criteria P1, P2, P3, P4, P5, P6, P7, P8, P9, P10; M1 and M2 and aspects of D1 and D2 from Unit 4 – Mathematics for Technicians.

Learning outcomes

1 Know how to use algebraic methods
2 Be able to use trigonometric methods and standard formulae to determine areas and volumes
3 Be able to use statistical methods to display data
4 Know how to use elementary calculus techniques

Content

1) Know how to use algebraic methods

Indices and logarithms: laws of indices ($a^m \times a^n = a^{m+n}$, $a^m/a^n = a^{m-n}$, $(a^m)^n = a^{mn}$), laws of logarithms ($\log A + \log B = \log AB$, $\log A^n = n \log A$, $\log A - \log B = \log (A/B)$, eg common logarithms (base 10), natural logarithms (base e), exponential growth and decay.

Linear equations and straight line graphs: linear equations, eg $y = mx + c$; straight line graph (coordinates on a pair of labelled Cartesian axes, positive or negative gradient, intercept, plot of a straight line); experimental data, eg Ohm's law, pair of simultaneous linear equations in two unknowns.

Factorisation and quadratics: multiply expressions in brackets by a number, symbol or by another expression in a bracket; by extraction of a common factor, eg $ax + ay$, $a(x + 2) + b(x + 2)$; by grouping, eg $ax - ay + bx + by$; quadratic expressions, eg $a^2 + 2ab + b^2$; roots of an equation, eg quadratic equations with real roots by factorisation, and by the use of formula.

2) Be able to use trigonometric methods and standard formulae to determine areas and volumes

Circular measure: radian; degree measure to radians and vice versa; angular rotations (multiples of π radians); problems involving areas and angles measured in radians; length of arc of a circle ($s = r\,\theta$); area of a sector ($A = \frac{1}{2}r^2\theta$).

Triangular measurement: functions (sine, cosine and tangent); sine/cosine wave over one complete cycle; graph of tan$A$ as A varies from 0° to 360°; (tan$A$ = sin A/ cos A); values of the trigonometric ratios for angles between 0° and 360°; periodic properties of the trigonometric functions; the sine and cosine rule; practical problems, eg calculation of the phasor sum of two alternating currents, resolution of forces for a vector diagram.

Mensuration: standard formulae to solve surface areas and volumes of regular solids, eg volume of a cylinder $= \pi r^2 h$, total surface area of a cylinder $= 2(\pi r^2) + 2\pi rh$, volume of a sphere $= 4/3\ \pi r^3$, surface area of a sphere $= 4\pi r^2$, volume of a cone $= 1/3\ \pi r^2 h$, curved surface area of a cone $= \pi r \times$ *slant height.*

3) Be able to use statistical methods to display data

Data handling: data represented by statistical diagrams, eg bar charts, pie charts, frequency distributions, class boundaries and class width, frequency table; variables (discrete and continuous); histogram (continuous and discrete variants); cumulative frequency curves.

Statistical measurements: arithmetic mean; median; mode; discrete and grouped data.

4) Know how to use elementary calculus techniques

Differentiation: differential coefficient; gradient of a curve $y = f(x)$; rate of change; Leibnitz notation (dy/dx); differentiation of simple polynomial functions, exponential functions and sinusoidal functions; problems involving evaluation, eg gradient at a point.

Integration: integration as the reverse of differentiating, basic rules for simple polynomial functions, exponential functions and sinusoidal functions; indefinite integrals; constant of integration; definite integrals; limits, evaluation of simple polynomial functions; area under a curve, eg $y = x(x - 3)$, $y = x^2 + x + 4$.

Grading criteria

P1 manipulate and simplify three algebraic expressions using the laws of indices and two using the laws of logarithms

The laws of indices become very useful when we are evaluating formulae containing quantities that are multiples and sub-multiples of 10. This often happens when we are dealing with units such as GN (N × 10^9), MW (W × 10^6), mA (A × 10^{-3}), μV (V × 10^{-6}), etc. When multiplying or dividing quantities such as these, you can often simplify the calculation by applying the laws of indices to reduce the different powers of 10 to one. The EXP key on your electronic calculator is used for entering powers of 10 and you will find it most useful when carrying out engineering calculations and deciding the units in which your answers are best expressed.

Logarithms are easily found by using the 'log' key on your electronic calculator. The logarithm of a number to the base 10, which is also known as a common logarithm, is the power that 10 must be raised by to give that number. It follows that the laws of logarithms are closely related to the laws of indices because when you add or subtract two logarithms you are adding or subtracting the powers of 10 that go to make up two numbers. The answer tells you the power of 10 that results when the numbers are multiplied together or divided and the second function of the 'log' key actually changes it to that number. In addition to logarithms to the base 10, we also have natural logarithms that are found by using the 'ln' key on your calculator. These have as their base the number e =2.718. They are particularly useful for analysing problems involving natural growth and decay, such as the growth and decay of current and voltage in electric circuits.

P2 solve a linear equation by plotting a straight-line graph using experimental data and use it to deduce the gradient, intercept and equation of the line

When you suspect that two variables in an engineering system are related in some way, it is usual to take readings during an experiment and plot them on a graph. A typical example is the distance moved by a vehicle in regular intervals of time. It is customary to plot the independent variable (in this case, time) on the horizontal axis and the dependant variable (velocity) on the vertical or ordinate axis. If the graph is a straight line, this immediately tells you that the two variables are directly proportional. Furthermore, the gradient of the graph gives you the constant of proportionality and the intercept on the ordinate axis tells you the value of the dependant variable when the independent variable is zero. When the variables increase together, the gradient is positive but when one increases and the other decreases by proportional amounts, the gradient is negative.

In algebra the independent variable is usually given as x and the dependant variable as y. The gradient or constant of proportionality is m, and the intercept on the y-axis is c. Once you have plotted a graph and found these values, you can write down the equation that connects the two variables, ie $y = mx + c$.

P3 factorise by extraction and grouping of a common factor from expressions with two, three and four terms respectively

When we multiply two or more numbers together, the answer is called the *product* and the numbers are said to be the *factors* of the product. The same applies to algebraic expressions and to achieve this assessment criterion you are required to factorise three such expressions. You will have to look carefully at them and decide whether you can obtain the factors by directly extracting a common factor or by first grouping the variables together and then bracketing their coefficients.

P4 solve circular and triangular measurement problems involving the use of radian, sine, cosine and tangent functions

You are required to do a number of things to achieve this assessment criterion. You must demonstrate an ability to convert degrees to radians and vice versa. You must also be able to express an angle as a multiple of π radians and calculate arc length and sector area for a given circle and angle subtended at its centre.

For triangular measurement you must be able to solve triangles making use of the sine, cosine and tangent functions.

P5 sketch each of three trigonometric functions over a complete cycle

To achieve this assessment criterion, you will need to draw up a table showing the values of sine, cosine and tangent for angles up to 360°. You should use your electronic calculator to obtain the values. Increments of 15° would be appropriate so that you can plot accurate graphs showing how the functions vary with increasing angle.

P6 produce answers to two practical engineering problems involving the sine and cosine rule

Here you will be asked to solve two engineering problems in which non-right-angle triangles occur. You might be asked to calculate the dimensions of a component. Alternatively you could be asked to find the sum of two vector quantities such as forces, or two phasor quantities such as alternating voltages or currents. You will need to examine the information you are given and then decide which of the rules to use first. Sometimes you only need one of them but very often you will need to use them both.

P7 use standard formulae to find surface areas and volumes of regular solids for three different examples respectively

Engineering installations such as tanks, boilers, pressure vessels, hoppers etc are often cylindrical, spherical or conical in shape. These shapes are known as *regular solids* and designers use the standard formulae for calculating their volume and surface area. These formulae are given in the Content section on p. 62. You may be asked to calculate the surface area and volume of a container that is a combination of two or more of these shapes. In such a case you will need to consider the two sections separately and them add their areas and volumes together, making full use of the x^2 and x^y function keys on your electronic calculator. The x^y key is useful for entering the r^3 term when calculating the volume of a sphere. On the more modern calculators this key is marked with a symbol that looks rather like a pitched roof.

P8 collect data and produce statistical diagrams, histograms and frequency curves

To achieve this assessment criterion, you could collect data from inspection activities or by carrying out a survey of product output. You could also carry out a survey of the number of people working in different occupations in an engineering company. The data then needs to be analysed and displayed graphically in the most appropriate way. A pie chart might be a good way of showing the relative numbers of people working in design, production, sales etc within a company. A bar chart might be a good way of showing weekly output figures over a period of two or three months, and a histogram or frequency curve might be a good way of displaying inspection data such as dimensional variation within a batch of components.

P9 determine the mean, median and mode for two statistical problems and explain the relevance of each average as a measure of central tendency

Mean, median and mode are all ways of measuring *central tendency* in a set of data. To achieve this assessment criterion, you will need to analyse two sets of data and find the mean, median and mode. You will then need to say which you think gives the best indication of central tendency for each set of data, giving reasons for your choice.

P10 apply the basic rules of calculus arithmetic to solve three different types of function by differentiation and two different types of function by integration

The three different types of function that you will need to differentiate to achieve this assessment criterion are described in the Content section on pp. 62–3. One will be a simple polynomial function containing powers and multiples such as $y = 2x^3 + 4x^2 + 3$. Another will be sinusoidal such as $y = 3\sin x - 5\cos x$, and the third one will be an exponential function such as $y = 6e^{2x} + 2e^{-x}$. Differentiating gives you an expression for finding the gradient of the graph of a function and you will need to calculate its value at a given value of x or y. To achieve this assessment criterion, you will also need to integrate a polynomial function and one of the other two types. When you write down the indefinite integrals, don't forget to include the constant of integration. The polynomial function will have limits and you will be required to calculate its definite integral, which is the area under its curve between those limits.

M1 solve a pair of simultaneous linear equations in two unknowns

The solution of a pair of simultaneous linear equations is the value of the two variables, which might be x and y, that satisfies both equations. There are three ways of doing this. One is to draw up a table of values and plot graphs of the two variables on the same set of axes. The graphs will both be straight lines and where they cross gives you the values of x and y, that satisfies both equations.

Another method is by *elimination* of one of the variables. To do this you have to multiply both sides of one equation by a number that makes the coefficient of one of its variables, say x, numerically equal to that of x in the other equation. You can then subtract or add the two equations together to eliminate x. This leaves an equation in which y is the only unknown and you can then find its value. Finally, you can put this y-value in either of the two equations to find the value of x that satisfies them both.

Another method is by *substitution* for one of the variables. To do this you take one of the equations and make one of the variables, say x, the subject of the equation. You then substitute this expression for x in the other equation. This again leaves y as the one unknown and you can find its value. You then put this back into the expression for x to find its value that satisfies both equations.

M2 solve one quadratic equation by factorisation and one by the formula method

When we use mathematical techniques to investigate engineering systems, we sometimes find that we have to solve a quadratic equation. The easiest ones to solve are those that can be manipulated so that coefficient of x^2 is 1, and the coefficient of x and the remaining constant in the equation are whole numbers. It is then sometimes possible to solve the equation by factorisation. To do this you have to try and find two numbers that when added together give the coefficient of x, and when multiplied together give the value of the remaining

constant. These numbers with their signs reversed are the values of x that satisfy the equation. They are also called the *roots* of the equation. Sometimes the above manipulation is not possible, but with practice you may still be able to spot the factors of the quadratic if the coefficients of x^2, x and the remaining constant are whole numbers.

If a quadratic equation proves too difficult to factorise or if the coefficients have decimal places, the alternative method of solution is by use of the quadratic solution formula. Here the x^2 and square root functions on your electronic calculator become very useful when evaluating the two possible values of x that are the roots of the equation.

D1 apply graphical methods to the solution of two engineering problems involving exponential growth and decay, analysing the solutions using calculus

Here you will be given two sets of experimental data to analyse. In both problems the independent variable will probably be time, and you will be asked to find the equation connecting the variables, and the rate of change of the dependant variable at some given instant. To achieve the assessment criterion, you will need to plot a graph for each problem, carefully draw a tangent to the curve at the required point and find its gradient by constructing a right-angle triangle of suitable size. Your next task will be to find the exponential growth or decay equation that connects the variables. This can be done by plotting the natural logarithm of the dependant variable against time. The result will be a straight line graph from whose gradient and intercept you can construct the equation. You can now check the answer you got for rate of change by differentiating the expression and finding the value of the differential at the instant required.

D2 apply the rules for definite integration to two engineering problems that involve summation

Definite integrals are ones that are evaluated between given limits to find the area under the graph that connects two given variables. Here you will be given the equations that connect the variables in two different engineering problems. Typical examples might be velocity and time, $v = f(t)$, force and distance, $F = f(s)$, or electric current and time, $I = f(t)$. To achieve the assessment criterion, you will be asked to find the area under the graphs that connects the variables. In the case of a velocity v time graph, the area will be the distance travelled over a given interval of time. In the case of the force v distance graph, the area under the curve will be the work done in moving through a given distance and in the case of the current v time graph, it will be the electric charge that has passed in a given time. You will need to find these areas by integrating the given expressions and evaluating the definite integrals between given limits.

ACTIVITY 1

Simplify the following expressions using the laws of indices and logarithms.

i) $\dfrac{2(x^3)^2 \times 3(x^2)^4}{4(x^4)^3}$

ii) $\dfrac{7(x^2)^6 \times 6(x^3)^2}{3(x^2)^3 \times 2(x^2)^4}$

iii) $3\log a^3 + 2\log a^4 - 6\log a^2$

iv) $\dfrac{4\log b^2 - 2\log b^3}{5\log b^4}$

Before attempting these, you might like to look at the following solutions to similar problems.

1 Simplify the expression $\dfrac{4(x^3)^2 \times 5(x^2)^3 \times 2(x^4)^2}{8(x^3)^6}$

Removing the brackets gives: $\dfrac{4x^6 \times 5x^6 \times 2x^8}{8x^{18}}$

Simplifying the numerator gives: $\dfrac{40x^{20}}{8x^{18}}$

This finally cancels down to: $5x^2$

2 Simplify the expression $4\log b^2 - 3\log b^3 + 2\log b^5$

The expression reduces to: $8\log b - 9\log b + 10\log b$

Adding and subtracting finally gives: $9\log b$

ACTIVITY 2

Task 1

The following are experimental readings of velocity and time for a moving body:

Time (t seconds)	2	4	6	8	10	12	14
Velocity (v m s^{-1})	4.5	5.7	6.9	8.1	9.3	10.5	11.7

Plot the graph of velocity (vertical axis) against time (horizontal axis) and obtain from it the equation that connects the two quantities.

Task 2

Solve the following sets of simultaneous equations using an analytical method and then check your answer by plotting their graphs on the same set of axes.

i) $4y = x + 4$

$2y = 2 - 3x$

ii) $3y = x - 2$

$y = 2x + 1$

If you are not sure how to solve the above equations, you might like to look at the following solutions to a similar problem.

Solve the simultaneous equations

$4y = 2x + 2$ (i)

$3y = 4x - 6$ (ii)

Method 1, by elimination: multiply (i) by 2 so that the coefficients of x are the same and then subtract as follows:

$$8y = 4x + 4$$
$$\underline{3y = 4x - 6}$$
$$5y = 0 + 10$$

This gives: $y = \dfrac{10}{5} = 2.$

Put this value of y into equation (ii) to find the value of x that satisfies them both.

ie $3 \times 2 = 4x - 6$

$6 = 4x - 6$

$12 = 4x$

This gives: $x = \dfrac{12}{4} = 3.$

Method 2, by substitution: transpose (i) to make x the subject

ie $4y - 2 = 2x$

$\dfrac{4y - 2}{2} = x$

This gives: $x = 2y - 1$

Now substitute for x in equation (ii)

ie $3y = 4(2y - 1) - 6$

$3y = 8y - 4 - 6$

$3y = 8y - 10$

$10 = 8y - 3y$

$10 = 5y$

This again gives: $y = \dfrac{10}{5} = 2$

Now put this value into the expression we obtained for x

This again gives: $x = (2 \times 2) - 1 = 3.$

ACTIVITY 3

Task 1

Factorise the following expressions by extraction and grouping of common factors.

i) $a(x + 2) + b(x + 2) + ax + bx$

ii) $3ax + 2b(x - 1) + 2bx + 3a(x - 1)$

Task 2

Find the roots of the following quadratic equations by factorisation and then checking your answer using the quadratic solution formula.

i) $x^2 + 4x - 5 = 0$

ii) $2x^2 + 2x - 24 = 0$

The following solutions to similar problems might help you with these tasks.

1 Factorise the expression $2ax + 2a(x - 1) + bx + b(x - 1)$

Rearranging gives: $2ax + bx + 2a(x - 1) + b(x - 1)$

Factorising gives: $x(2a + b) + (x - 1)(2a + b)$

Taking out the common factor gives: $(2a + b)(x + x - 1)$

Which finally becomes: $(2a + b)(2x - 1)$

2 Find the roots of the quadratic equation $3x^2 + 9x - 12 = 0$

Dividing throughout by 3 gives: $x^2 + 3x - 4 = 0$

This factorises into: $(x + 4)(x - 1) = 0$

The roots of the equation are thus $x = 1$ and $x = -4$

Checking with the quadratic formula,

$$x = \frac{-b \pm \sqrt{b^2 - 4ac}}{2a}$$

$$x = \frac{-9 \pm \sqrt{(9^2 - \{4 \times 3x - 12\})}}{2 \times 3}$$

$$x = \frac{-9 \pm \sqrt{(225)}}{6}$$

$$x = \frac{-9 \pm 15}{6}$$

this again gives, $x = 1$ and $x = -4$.

ACTIVITY 4

For the circle shown above you are required to find a) the length of the arc *AB*, b) the length of the chord *AB*, c) the area of the sector *OAB*, d) the area of the triangle *OAB*.

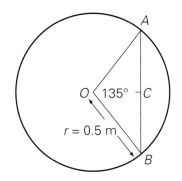

First you will need to convert the angle subtended at the centre of the circle into radians using the formula,

$$\theta \text{ rad} = \theta° \times \frac{2\pi}{360}$$

The length of the arc *AB* can then be found using the formula,

Arc length AB $= r\theta$ (where θ is measured in radians)

You will need to use trigonometry to find the length of the chord *AB*

Length AB $= 2 \times r\sin\theta/2$

The area of the sector *OAB* is found using the formula,

Area of sector OAB $= \frac{1}{2}r^2\theta$ (where θ is again measured in radians)

Before you can find the area of the triangle *OAB*, you will need to find the distance *OC*, which is the perpendicular distance from *AB* to the centre of the circle.

ie *OC* $= r\cos\theta/2$

You can now find the area of triangle *OAB* using the formula,

Area of triangle OAB $= \frac{1}{2} \times AB \times OC$

Note: Your electronic calculator can be programmed in degree or radians mode. If you decide to work in radians throughout, make sure that it is programmed in radians mode before you enter the angle $\theta/2$ to find sine and cosine values.

ACTIVITY 5

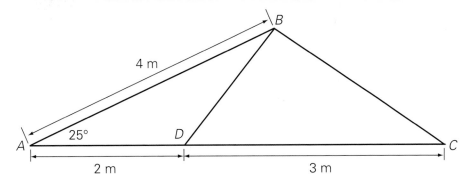

Find the lengths of the members *BC* and *BD* in the roof truss shown above.

69

You will need to apply the cosine rule to find the length of member BC

ie $BC^2 = AB^2 + AC^2 - 2.AB.AC.\cos A$

or $BC = \sqrt{(AB^2 + AC^2 - 2.AB.AC.\cos A)}$.

You will then need to apply the sine rule to find the angle at C

ie $\dfrac{BC}{\sin A} = \dfrac{AB}{\sin C}$

or $\sin C = \dfrac{AB}{BC} \sin A$.

Having found the angle at C, it can now be used in the cosine rule to find the length of the member BD

ie $BD^2 = BC^2 + DC^2 - 2.BC.DC.\cos C$

or $BD = \sqrt{(BC^2 + DC^2 - 2.BC.DC.\cos C)}$.

ACTIVITY 6

Calculate the surface area and the volume contained by the mixer drum shown.

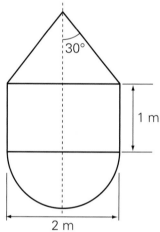

There are three definite parts to this fabrication, the centre cylindrical portion being the easiest to deal with using the following formulae.

Surface area of cylindrical section $= 2\pi rl$ or πdl (where $l =$ the height of the cylinder)

Volume of cylindrical section $= \pi r^2 l$ or $\dfrac{\pi d^2 l}{4}$

The formulae for hemispherical end section are as follows.

Surface area of hemispherical section $= \frac{1}{2} \times 4\pi r^2$ or $2\pi r^2$

Volume of hemispherical section $= \frac{1}{2} \times \dfrac{4\pi r^3}{3}$ or $\dfrac{2\pi r^3}{3}$

Before you can find the surface area and volume of the conical section you will need to calculate its slant height and vertical height.

Slant height of conical section, $h_s = \dfrac{r}{\sin 30}$

Vertical height of conical section, $h_v = \dfrac{r}{\tan 30}$

The surface area and volume of the conical section can now be found using the following formulae.

Surface area of conical section $= \pi r h_s$

$$\textit{Volume of conical section} = \frac{\pi r^2 h_v}{3}$$

Finally, you can add together the separate surface areas and separate volumes to find the total surface area and total volume of the drum.

ACTIVITY 7

Task 1

An engineering company employs the following categories of personnel. You are required to display this information in the form of a pie chart.

Design	*30*
Production	*545*
Purchasing	*50*
Administration	*65*
Sales	*18*
Maintenance	*12*

Task 2

The company profits for the seven years from 2000 to 2006 are as follows. You are required to display this information in the form of a bar chart.

2000 – £10.5 million
2001 – £11.3 million
2002 – £13.8 million
2003 – £12.5 million
2004 – £14.1 million
2005 – £15.5 million
2006 – £13.5 million

The first thing that you need to do is find the total number of employees in the company (see Task 1). You then have to calculate the angles on your pie chart that will represent each class of employee. For the design personnel the angle will be:

$$\textit{Design personnel} = \left[\frac{30}{\textit{Total number of employees}} \right] \times 360°$$

You will need to repeat this calculation for each class of employee and then check that the angles add up to 360°. Having decided on a suitable diameter for your pie chart, you can then proceed to construct it using compasses and a protractor.

In Task 2 you will need to decide on a suitable scale for your bar chart. For instance, you may decide to use 10 mm = £1 million with a bar width of 10 mm. You can then proceed to calculate the length of each bar and construct the bar chart. You will probably get the best results if you use graph paper for the task.

ACTIVITY 8

Task 1

The masses of a batch of 50 castings produced by the company are recorded below, correct to the nearest 0.1 kg. You are required to draw up a tally chart with 9 classes of class width 0.3 kg and then present the data in the form of a) a frequency polygon, b) a histogram.

6.8	7.2	6.6	7.5	8.1	7.0	6.5	7.2	7.6	7.0
7.3	6.1	7.1	7.3	7.7	6.8	7.7	7.5	7.4	7.5
6.5	6.7	7.2	6.8	7.1	6.2	7.8	6.9	7.4	6.7
7.1	6.4	7.8	7.0	7.4	7.5	7.1	6.3	8.0	7.6
7.0	6.9	7.5	6.9	7.2	7.3	6.7	7.4	7.2	6.4

The lowest class for the castings will be *6.1* to *6.3* kg in which there are 3 castings. A possible format for your tally chart in Task 1 is as follows.

Class	Tally	
5.75 to 6.05	–	(0)
6.05 to 6.35	III	(3)
6.35 to 6.65	₩	(5)
6.65 to 6.95	etc	
6.95 to 7.25		
7.25 to 7.55		
7.55 to 7.85		
7.85 to 8.15		
8.15 to 8.45	–	(0)

Your frequency polygon will be a graph plotted of the number in each class (vertical axis) against the mid-point value of each class (horizontal axis). The points should be joined up by a series of straight lines, starting at zero for the mid-point of class *5.75* to *6.05* kg in which there are no castings, and ending at zero for the mid-point of class *8.15* to *8.45* kg in which there are again no castings.

Your histogram, which contains the same information, will be a series of vertical blocks whose width is the class width and whose height is the number or 'frequency' in each class. You should use graph paper, preferably in 'landscape' orientation for best results, and choose scales that make full use of each sheet.

Task 2

The times in hours to complete a maintenance operation in the company, on nine successive occasions, are as follows. You are required to determine the mean, median and mode of the set of data and say which you think might give the best indication of central tendency.

| 31 | 29 | 22 | 26 | 30 | 31 | 22 | 32 | 20 |

You will need to rearrange the values in ascending order, from which you should immediately be able to identify the median, ie the value at the centre of the set. You should find that there are two modal values from which you might decide that in this case the mode is not a good way of indicating central tendency. To find the mean, you will need to use the following formula.

$$\text{Arithmetic mean value} = \frac{\text{Total number of hours}}{\text{Number of maintenance operations}}$$

Finally, you should examine the mean, median and modal values and decide which you think gives the best indication of how long the maintenance activity might typically be expected to take.

ACTIVITY 9

Determine the differential coefficients of the following functions.

i) $y = 3x^4 - 2x^3 + 5x + 1$

ii) $y = 5\sin 3x - 2\cos 4x$

iii) $y = 3e^{2x} + 4e^{-3x}$

The following solutions to similar problems might help you with these tasks.

Evaluate the differential coefficient of the function
$y = 2x^5 + 4x^3 - 3x + 2$

$$y = 2x^5 + 4x^3 - 3x + 2$$

$$\frac{dy}{dx} = 10x^4 + 12x^2 - 3$$

Evaluate the differential coefficient of the function
$y = 4\cos 2x + 3\sin 4x$

$$y = 4\cos 2x + 3\sin 4x$$

$$\frac{dy}{dx} = -8\sin 2x + 12\cos 4x$$

Evaluate the differential coefficient of the expression $y = 4e^{3x} + 2e^{-2x}$

$$y = 4e^{3x} + 2e^{-2x}$$

$$\frac{dy}{dx} = 12e^{3x} - 4e^{-2x}$$

ACTIVITY 10

The velocity v, of a body in ms^{-1} at an instant in time t, is given by the equation

$$v = 2t^2 + 4t + 3,$$

where the time, t, is in seconds. Given that the area under the velocity–time graph is the distance travelled in metres, you are required to find the distance travelled in the time interval between $t = 1$ second and $t = 3$ seconds.

The following solution to a similar problem might help you with this task.

The force F (Newtons), acting on a body is related to distance travelled s (metres), by the equation

$F = 4s^2 + 3s + 2$

Find the work done on the body whilst travelling between the points where $s = 2$ m and $s = 4$ m.

The work done is the area under the force versus distance graph between the given limits, which we can find as follows by integration.

$W = \int F\, ds$

$$W = \int_{2}^{4} (4s^2 + 3s + 2)\, ds$$

$$W = \left[\frac{4s^3}{3} + \frac{3s^2}{2} + 2s \right]_{2}^{4}$$

$$W = \left[\frac{(4 \times 4^3)}{3} + \frac{(3 \times 4^2)}{2} + (2 \times 4) \right] - \left[\frac{(4 \times 2^3)}{3} + \frac{(3 \times 2^2)}{2} + (2 \times 2) \right]$$

$W = 117.3 - 20.7$

$W = 96.6$ N

ANSWERS

ACTIVITY 1

i) $1.5x^2$

ii) $7x^4$

iii) $5\log a$

iv) 0.1

ACTIVITY 2

Task 1

$v = 0.6t + 3.3$

Task 2

i) $y = 1, x = 0$

ii) $y = -1, x = -1$

ACTIVITY 3

Task 1

i) $(a + b)(2x + 2)$

ii) $(3a + 2b)(2x - 1)$

Task 2

i) $x = 1$ and $x = -5$

ii) $x = 3$ and $x = -4$

ACTIVITY 4

1.18 m, 0.92 m, 0.29 m^2, 0.09 m^2

ACTIVITY 5

$BC = 2.18$ m and $BD = 2.34$ m

ACTIVITY 6

18.85 m^2 and 7.05 m^3

ACTIVITY 10

39.3 m

This section focuses on grading criteria P1, P2, P3, P4, P5, P6, P7, P8, P9, P10; M1, M2, M3 and aspects of D1 and D2 from Unit 5 – Electrical and Electronic Principles.

Learning outcomes

1 Be able to use circuit theory to determine voltage, current and resistance in direct current (DC) circuits

2 Understand the concepts of capacitance and determine capacitance values in DC circuits

3 Understand the principles and properties of magnetism

4 Understand single-phase alternating current (AC) theory

Content

1) Be able to use circuit theory to determine voltage, current and resistance in direct current (DC) circuits

DC circuit theory: voltage, eg potential difference, electromotive force (emf); resistance, eg conductors and insulators, resistivity, temperature coefficient, internal resistance of a DC source; circuit components (power source, eg cell, battery, stabilised power supply; resistors, eg function, types, values, colour coding; diodes, eg types, characteristics, forward and reverse bias modes); circuit layout (DC power source, resistors in series, resistors in parallel, series and parallel combinations); Ohm's law, power and energy formulae, eg $V = IR$, $P = IV$, $W = Pt$, application of Kirchhoff's voltage and current laws.

DC networks: networks with one DC power source and at least five components, eg DC power source with two series resistor and three parallel resistors connected in a series parallel arrangement; diode resistor circuit with DC power source, series resistors and diodes.

Measurements in DC circuits: safe use of a multimeter, eg setting, handling, health and safety; measurements (circuit current, voltage, resistance, internal resistance of a DC power source, testing a diode's forward and reverse bias).

2) Understand the concepts of capacitance and determine capacitance values in DC circuits

Capacitors: types (electrolytic, mica, plastic, paper, ceramic, fixed and variable capacitors), typical capacitance values and construction (plates, dielectric materials and strength, flux density, permittivity); function, eg energy stored, circuits (series, parallel, combination); working voltage.

Charging and discharging of a capacitor: measurement of voltage, current and time; tabulation of data and graphical representation of results; time constants.

DC network that includes a capacitor: eg DC power source with two/three capacitors connected in series, DC power source with two/three capacitors connected in parallel.

3) Understand the principles and properties of magnetism

Magnetic field: magnetic field patterns, eg flux, flux density (B), magnetomotive force (mmf) and field strength (H), permeability, B/H curves and loops; ferromagnetic materials; reluctance; magnetic screening; hysteresis.

Electromagnetic induction: principles, eg induced electromotive force (emf), eddy currents, self and mutual inductance; applications (electric motor/generator, eg series and shunt motor/generator; transformer, eg primary and secondary current and voltage ratios); application of Faraday's and Lenz's laws.

4) Understand single-phase alternating current (AC) theory

Single phase AC circuit theory: waveform characteristics, eg sinusoidal and non-sinusoidal waveforms, amplitude, period time, frequency, instantaneous, peak/peak-to-peak, root mean square (rms), average values, form factor; determination of values using phasor and algebraic representation of alternating quantities, eg graphical and phasor addition of two sinusoidal voltages, reactance and impedance of pure R, L and C components

AC circuit measurements: safe use of an oscilloscope, eg setting, handling, health and safety; measurements (periodic time, frequency, amplitude, peak/peak-to-peak, rms and average values); circuits, eg half and full wave rectifiers

Grading criteria

P1 use DC circuit theory to calculate current, voltage and resistance in DC networks

You need to be able to apply Ohm's law ($V = IR$) in order to determine the voltage, current and resistance in simple direct current (DC) circuits. You need to know about voltage and the difference between the potential difference (voltage drop) that appears across a component (such as a resistor) and the electromotive force (emf) that is provided by a battery. You need to understand the difference between conductors and insulators and you need to know about resistors and resistance. You should be able to calculate the resistance of a conductor given its physical dimensions and resistivity (or specific resistance). You also need to know how temperature affects resistance and you should be able to calculate the value of a resistor given its temperature and the temperature coefficient of resistance.

All sources of direct current (DC) have some internal resistance. You need to understand what effect this has on the voltage of a DC source as current when it supplies current to a load. You need to know about the various different types of component that make up a simple DC circuit. You should recognise the symbols used for the cells, batteries, power supplies and resistors and you should be able to draw simple circuits that use these components.

Resistors are often marked with a colour code that indicates their value. You need to know the system used for colour coding and be able to determine the value and tolerance of a resistor from its markings.

P2 use a multimeter to carry-out circuit measurements in a DC network

A multimeter is an electrical test instrument used for measuring voltage, current and resistance in a circuit. You need to be able to use such an instrument to make measurements of voltage, current and resistance in a DC network. You also need to be able to demonstrate that you are using the instrument in a safe way. This means using the correct test leads and probes and making appropriate connections to the circuit under test for each measurement that you make. It also means setting the correct function and range before use and ensuring that you comply with relevant health and safety precautions.

Typical measurements that you might have to make include supply and branch current, supply and nodal voltage, resistance, internal resistance of a DC power source and testing a diode using both forward and reverse bias.

P3 compare the forward and reverse characteristics of two different types of semi-conductor diode

Semi-conductors are an important class of materials used in the manufacture of a variety of useful electronic components such as diodes and transistors. Diodes are able to pass current in one direction but not in the other. Thus they act as conductors or insulators according to how they are connected. This is a useful property which you need to understand. You also need to know the main types of diode and how their characteristics differ.

P4 describe the types and function of capacitors

You need to be able to identify various different types of capacitor types (electrolytic, mica, plastic, paper, ceramic, fixed and variable capacitors). For each type you need to be able to specify typical capacitance values and construction (plates, dielectric materials, electric field strength, and permittivity).

You need to know what a capacitor does and what it is used for. You also need to identify the symbol used for different types of capacitor (eg fixed, variable, electrolytic).

P5 carry out an experiment to determine the relationship between the voltage and current for a charging and discharging capacitor

77

You need to carry out an experiment to find out what happens to the voltage and current in a circuit when a capacitor is being charged (ie when charge is being placed in it) and when it is being discharged (ie when charge is being removed from it). Your experiment needs to involve measurements of voltage, current and time which can be tabulated and used to construct a graph showing how the voltage/current in the circuit varies with time.

You also need to be able to relate the values of capacitance and resistance used in your experiment to the time constant of the circuit (ie the product of the C and R values).

P6 calculate the charge, voltage and energy values in a DC network that includes a capacitor

You need to be able to calculate the charge and energy stored in a capacitor given its capacitance and the voltage that appears across the plates.

P7 describe the characteristics of a magnetic field and explain the relationship between flux density (B) and field strength (H)

A magnetic field appears in the space that surrounds a conductor when it carries an electric current. You need to know that the strength of the field depends on the shape of the conductor and the current flowing in it. You also need to know that the strength of the magnetic field is a measure of the density of the flux at any particular point. You also need to know the effect of the shape of the conductor (whether it is straight or formed into a loop or coil) and the material through which the flux passes.

You need to know how the field strength depends on the magnetomotive force (mmf) and you need to understand the relationship that exists between the flux (Φ), flux density (B), field strength (H). You also need to know how the ability of a material to support a magnetic flux (ie its permeability, μ) varies with the applied field strength. You should be able to construct graphs showing this relationship in the form of a B–H curve or loop.

You must also be able to identify common ferromagnetic materials (such as cast iron, mild steel, carbon steel, etc) and their properties, characteristics and applications.

P8 describe the principles and applications of electromagnetic induction

You need to know that an inductor is a conductor wound into the form of a coil and this may or may not have a core made from a ferromagnetic material which serves to concentrate the flux. You also need to know that an emf will be generated across the ends of a coil whenever the current flowing in it changes. You need to explain that this is due to a property known as self-inductance. This effect has a number of applications and you need to be aware of them. You also need to know that, because they are conductors, eddy currents can be induced in ferromagnetic cores. In addition, you need to understand that mutual inductance exists when two coils are coupled together or when they share the same ferromagnetic core.

An important application of electromagnetic induction is the transformer. You need to be able to explain how a transformer operates and you should be able to calculate the primary and secondary currents and voltages when given the number of turns present on the primary (input) and secondary (output) windings.

P9 use single phase AC circuit theory to explain and determine the characteristics of a sinusoidal AC waveform

Alternating currents are currents that continuously reverse their direction of flow. The polarity of the voltage produced when an alternating current flows must consequently also be changing continuously from positive to negative and vice versa. You need to be able to describe an alternating current and you must be able to sketch a typical sinusoidal voltage or current, indicating the axes of voltage and time. You need to know the relationship between frequency and periodic time and be able to use this to calculate typical values. You also need to know the relationship between average, peak, peak-to-peak and root mean square (rms) values for a sine wave and you should be able to apply this to solve simple problems.

P10 use an oscilloscope to measure and determine the inputs and outputs of a single phase AC circuit

You need to be able to make measurements on a simple AC circuit, displaying input and output voltage waveforms to a common time scale using an oscilloscope. You should be able to use the oscilloscope to measure the peak and peak-to-peak voltage of the waveform and you should be able to measure the time difference and phase angle between two waveforms.

When alternating voltages are applied to capacitors or inductors, the magnitude of the current flowing will depend upon the value of capacitance or inductance and on the frequency of the current. In effect, capacitance and inductance oppose the flow of current in much the same way as a resistor. The important difference being that the effective resistance (we call this *reactance*) of the component varies with frequency (unlike the case of a resistor where the magnitude of the current does not change with frequency).

You need to be able to apply the formulae associated with reactance in order to calculate the reactance of a capacitor or an inductor at a given frequency. This will allow you to determine the current and voltage in an AC circuit that contains a capacitor or an inductor.

M1 use Kirchhoff's laws to determine the current in all the branches of a network containing two voltage sources, five nodes and power dissipated in a load resistor

Practical electrical and electronic circuits can contain many electronic components. These are connected to form series and/or parallel networks. You need to understand how these behave in terms of the voltages and current that are applied to them and you need to be able to determine the resistance of a network in which several resistors are present connected in both series and parallel. In order to calculate the voltages and currents in a network, you need to know (and be able to apply) Kirchhoff's current law and Kirchhoff's voltage law. You also need to determine the power and energy in a DC circuit by applying appropriate formulae such as $P = IV$ and $W = Pt$.

M2 evaluate capacitance, charge, voltage and energy in a network containing a series-parallel combination of three capacitors

You need to find the value of capacitance, charge, voltage and energy in a series/parallel network of capacitors and use the relationships that exist between these quantities.

M3 compare the results of adding and subtracting two sinusoidal AC waveforms graphically and by phasor diagram

You need to be able to add to and subtract from sinusoidal waveforms (both voltage and current) by constructing a graph and by using a phasor diagram. The graphical method requires that you plot each waveform separately and then add together voltage values at set time intervals in order to produce a third waveform which you then plot. The phasor method requires that you draw lines to represent the magnitude and relative phase angle of the two voltages or currents and use these to construct a resultant from a parallelogram of forces. The length of the resultant indicates its magnitude (ie amplitude or peak value) whilst its angle relative to the reference phasor (usually the supply current or supply voltage) gives its *phase angle*.

D1 analyse the operation and the effects of varying component parameters of a power supply circuit that includes a transformer, diodes and capacitors

You need to be able to make measurements in AC circuits using a multimeter (to measure rms values of voltage and current) and an oscilloscope (to measure peak and peak-to-peak values of voltage and current and also to observe the shape of a waveform and be able to measure periodic time and frequency).

You will need to be able to investigate a power supply circuit; measuring the voltages, currents and waveforms present using a multimeter and an oscilloscope and then interpreting the results of your measurements under different load conditions and also when using a range of different component values.

You also need to be able to show that you are using these test instruments in a safe way. This means using the correct test leads and probes and making appropriate connections to the circuit under test for each measurement that you make. It also means setting the correct function and range before use and ensuring that you comply with relevant health and safety precautions.

D2 evaluate the performance of a motor and a generator by reference to electrical theory

Applications of electromagnetism include motors and generators. You need to know the basic construction and operation of motors and generators. You need to explain how the magnetic field is produced for motors and generators and you should be able to sketch circuits showing series and parallel (shunt) connected field windings.

ACTIVITY 1

Use DC circuit theory to calculate the voltage, current and resistance in the network shown in figure 1:

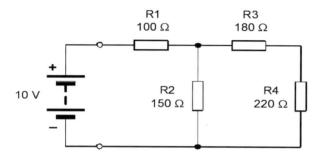

Figure 1

It's a good idea to start a problem like this by marking on the circuit all of the unknown currents and voltages that you are trying to find, as shown in figure 2:

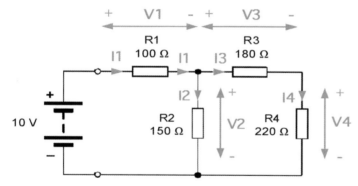

Figure 2 Unknown currents and voltages marked on the circuit shown in figure 1

You need to adopt a logical convention for numbering the currents and voltages. For example, I_1 is the current flowing in $R1$ and V_1 is the voltage dropped across it.

Next you should find the *effective* resistance of the network (ie the value of one single resistor that could replace all of the other resistors in the network). The effective resistance is the resistance that you would 'see' when looking into it from the supply terminals. You can do this in convenient stages, progressively reducing the number of resistors in the network by combining series and parallel resistors in each step, as shown below:

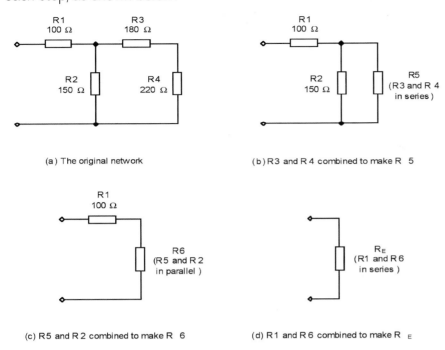

(a) The original network

(b) R3 and R4 combined to make R 5

(c) R5 and R2 combined to make R 6

(d) R1 and R6 combined to make R E

Figure 3 Stages in reducing the network to find its effective resistance

Having found the effective resistance of the network, you can now apply Ohm's law to calculate the current supplied to it from the 10 V supply. You can do this by using:

$$I = \frac{V}{R}$$

where $V = 10$ V and R is the value of R_E that you have calculated. The result that you have obtained from this calculation is the current supplied to the network and is also the current flowing in $R1$ (look back at figure 2 to confirm this). Since you now know I_1 you can apply Ohm's law again in order to determine the voltage drop across $R1$. You can do this by using:

$$V = IR$$

where R is the value of $R1$ (100 Ω) and I is the value of I_1 that you have already calculated.

Next you can apply Kirchhoff's voltage law to the network to determine the voltage dropped across $R2$, as follows:

$$10 = V_1 + V_2$$

or

$$V_2 = 10 - V_1$$

where V_1 is the voltage dropped across $R1$ that you calculated previously.

Now that you know V_2 you can use Ohm's law to calculate the current in $R2$. When you have this current you can apply Kirchhoff's current law to determine the current flowing in R3 (note that the same current flows in R4 as these two components are connected in series, thus $I_3 = I_4$). Applying Kirchhoff's current law at the junction of R1, R2 and R3 gives:

$I_1 = I_2 + I_3$

from which

$I_3 = I_1 - I_2$

Finally, you know that:

$I_3 = I_1$

Calculate V_3 and V_4 using Ohm's law.

It's a good idea to show all of your working in your answer and also to record your results in a table like the one shown below:

	R1	R2	R3	R4
Current flowing (mA)	Insert your calculated value for I_1	Insert your calculated value for I_2	Insert your calculated value for I_3	Insert your calculated value for I_4
Voltage drop (V)	Insert your calculated value for V_1	Insert your calculated value for V_2	Insert your calculated value for V_3	Insert your calculated value for V_4

ACTIVITY 2

Use a multimeter to carry out current and voltage measurements on the circuit shown in figure 1 (note that this is the same circuit that you used in Activity 1 so you should already have the calculated values of voltage and current for this circuit):

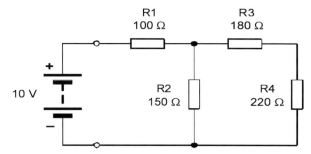

Figure 1

You should have noticed that there are several branches and nodes in this circuit. (A node is simply a common connection point in a circuit, usually where a number of components are connected together.)

This means that there will be quite a few voltages and currents that will need to be measured! The eight measuring points for the network are shown in figure 2. You will obviously not be able to make all of these measurements at the same time. Instead, you will have to move the multimeter around the circuit. Connect and adjust the multimeter for the required current or voltage measurement at each point.

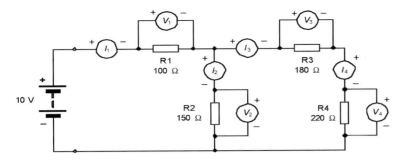

Figure 2 Measuring points for current and voltage in the network

Measuring the current

To measure the current flowing, you will need to set the multimeter to the correct DC current range before breaking the circuit at the point under investigation and insert the multimeter. We've shown this in figure 3 and how the meter is adjusted in figure 4.

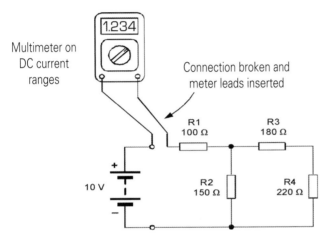

Figure 3 Multimeter connections for measuring the current in R1

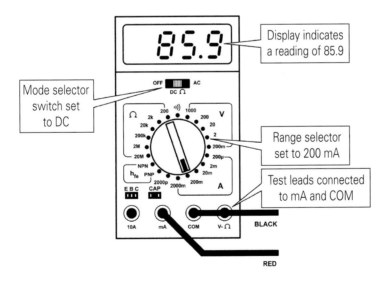

Figure 4 Multimeter adjusted for measuring DC current

Measuring the voltage

To measure the current flowing, you will need to set the multimeter to the correct DC voltage range before connecting the multimeter across the component under investigation. We've shown this in figure 5 and how the meter is adjusted in figure 6.

83

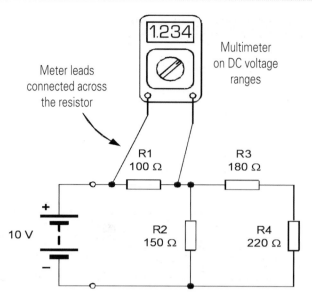

Figure 5 *Multimeter connections for measuring the voltage drop across R1*

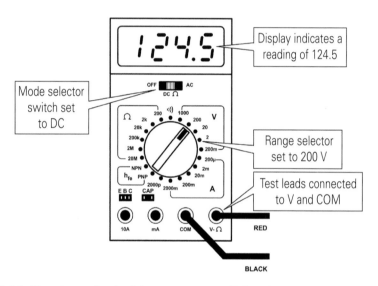

Figure 6 *Multimeter adjusted for measuring DC voltage*

As for Activity 1, it's a good idea to record your results in a table like the one shown below:

	R1	R2	R3	R4
Current flowing (mA)	Insert your measured value for I_1	Insert your measured value for I_2	Insert your measured value for I_3	Insert your measured value for I_4
Voltage drop (V)	Insert your measured value for V_1	Insert your measured value for V_2	Insert your measured value for V_3	Insert your measured value for V_4

Finally, it's worth comparing your measured results for the network with those that you obtained from calculation in Activity 1. If there are any significant differences it's important to suggest *why* these might have occurred!

ACTIVITY 3

Investigate a variety of different types of capacitor, including mica, ceramic, polystyrene, electrolytic and air-spaced variable types. For each type, identify typical construction, dielectric, capacitance range, tolerance, voltage rating, temperature coefficient and stability. Record your findings in a copy of the table shown below. The entry for mica capacitors has been included to get you started.

Property or feature	Capacitor type				
	Mica	Ceramic	Polystyrene	Electrolytic	Air-spaced variable
Construction	Multiple interleaved flat plates				
Dielectric material	Mica sheet				
Capacitance range	1 pF to 1 nF				
Typical tolerance	±1%				
Typical voltage rating (V)	350 V				
Temperature coefficient (ppm/°C)	+50				
Stability	Excellent				
Ambient temperature range (°C)	−40 to +125				
Typical applications	Tuned circuits and oscillators				

Use the information that you have obtained to answer the following questions:

1 Which type of capacitor is available with the largest capacitance values?

2 Which type of capacitor is available with the smallest capacitance values?

3 Which type of capacitor has the highest voltage ratings?

4 Which type of capacitor has the closest tolerance?

5 Which type of capacitor has the smallest size for a given value?

6 Which type of capacitor has the highest stability?

7 Which type of capacitor would be suitable for:

(i) tuning a radio receiver?

(ii) acting as a 'reservoir' capacitor in a power supply?

You will need to use a variety of different information sources such as books, electronic component catalogues, data sheets, and the Internet. Your school or college should be able to provide you with access to relevant documents.

ACTIVITY 4

Solve the following problems on capacitors:

1. A capacitor is required to store a charge of 40 µC when connected to a 50 V DC supply. Determine the value of capacitance required. Show all your working and state clearly the relationships on which you have based your calculations.

2. A capacitor of 32 µF is charged from a 150 V DC supply. Determine the charge in the capacitor and the energy stored in it. Show all your working and state clearly the relationships on which you have based your calculations.

To answer question 1, you will need to use the basic equation relating charge (Q), capacitance (C) and voltage (V) for a capacitor. You will find this in any electrical textbook (you may need to rearrange the equation to make C the subject).

To answer question 2, you will need to use the basic equation relating charge (Q), capacitance (C) and voltage (V), for a capacitor as well as the equation relating energy (W), capacitance (C) and voltage (V). Once again, you will find these in any electrical textbook.

ACTIVITY 5

The circuit in figure 1 shows a network of three capacitors.
Determine the following:

a) the potential difference that will appear across each capacitor

b) the charge present in each capacitor

c) the energy stored in each capacitor

d) the total energy stored in the circuit

e) the value, and working voltage, of one single capacitor that could replace the three capacitors.

Show all your working and state clearly the relationships on which you have based your calculations.

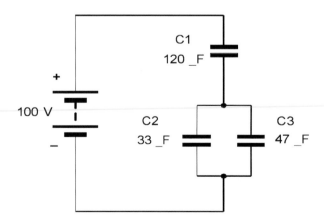

To answer this question, you will need to use both of the equations that you used in Activity 2. You will also need to decide which of the capacitors are connected in series and which are in parallel and reason out how the charge and voltage is shared between them. Note that the charge present in C_1 will be the same as that which is shared between C_2 and C_3.

You should start by determining the effective capacitance of the circuit. You can do this by finding the parallel combination of C_2 and C_3 (call this C_4) and then finding the series combination of C_1 and C_4. You will then be able to use the basic equation relating charge (Q), capacitance (C) and voltage (V), in order to determine the charge in the circuit. This charge will appear in C_1 and will also be shared between C_2 and C_3.

Having determined the charge, you will be able to calculate the voltage present in C_1 and (by applying Kirchhoff's voltage law) you will be able to find the voltage dropped across C_2 and C_3. Knowing these voltages will allow you to calculate the energy stored in each capacitor. The total energy will simply be the sum of the individual energies (you can also find this by using the effective capacitance of the circuit that you found earlier). Finally, you should make sure that you include all of your working in your answer.

ACTIVITY 6

Carry out an experiment to investigate the charge and discharge of a capacitor. You will need to carry out a number of measurements of voltage at regular time intervals as a capacitor is being charged and as it is being discharged. The charging current is to be supplied from a DC power supply via a large value resistor. The same resistor is to be used for discharging the capacitor. By using different values of resistor you will be able to investigate the behaviour of the circuit for different time constants (i.e. for different combinations of C and R).

Components and test equipment

Breadboard, 10 V DC power supply, digital multimeter with test leads, resistors of 100 kΩ, 220 kΩ and 47 kΩ, capacitor of 1000 μF, insulated wire links (various lengths), assorted crocodile leads, short lengths of black, red and green insulated solid wire. A watch or clock with a seconds display will also be required for timing.

Procedure

a) *Charging circuit*

Connect the charging circuit with $R = 100$ kΩ and $C = 1000$ μF, as shown in figure 1.

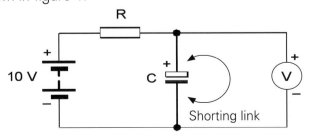

Figure 1 Charging circuit

You should start by placing a temporary shorting link across the capacitor. You should set the meter to the 20 V DC range and then remove the shorting link. After the link has been removed, you should measure and record the capacitor voltage at 25 s intervals over the range 0 to 250 s, recording your results in a table like the one shown below. You should repeat the measurements with $R = 220$ kΩ and $R = 47$ kΩ.

Charging circuit with $R = 100$ kΩ, $C = 1000$ µF

Time (s)	0	25	50	75	100	125	150	175	200	225	250
Voltage (V)											

Charging circuit with $R = 200$ kΩ, $C = 1000$ µF

Time (s)	0	25	50	75	100	125	150	175	200	225	250
Voltage (V)											

Charging circuit with $R = 47$ kΩ, $C = 1000$ µF

Time (s)	0	25	50	75	100	125	150	175	200	225	250
Voltage (V)											

b) *Discharging circuit*

Connect the discharging circuit with $R = 100$ kΩ and $C = 1000$ µF, as shown in figure 2.

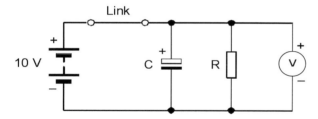

Figure 2 Discharging circuit

You should leave the link in place for a few seconds after the supply voltage has been switched on and then remove the link. You should then measure and record the capacitor voltage at 25 s intervals over the range 0 to 250 s from removing the link. Your results can be recorded in a table showing capacitor voltage against time. You should repeat the measurements with $R = 220$ kΩ and $R = 47$ kΩ.

Discharging circuit with $R = 100$ kΩ, $C = 1000$ µF

Time (s)	0	25	50	75	100	125	150	175	200	225	250
Voltage (V)											

Discharging circuit with $R = 220$ kΩ, $C = 1000$ µF

Time (s)	0	25	50	75	100	125	150	175	200	225	250
Voltage (V)											

Discharging circuit with R = 47 kΩ, C = 1000 µF

Time (s)	0	25	50	75	100	125	150	175	200	225	250
Voltage (V)											

Graphs and calculations

Use the graph paper to plot graphs of voltage (on the vertical axis) against time (on the horizontal axis) for both the charging and discharging circuits (and for each $C-R$ combination).

For the circuit with $R = 220$ kΩ, $C = 1000$ µF, you should use Ohm's law to calculate the charging current and discharging current at each time. You can then add the current graphs to the voltage graphs that you previously plotted.

Finally, calculate the time constant for each combination of resistance and capacitance that you used in the investigation. You can then mark the time constant on the time scale of each graph. The time constant for a circuit is given by $T = C \times R$.

You should tabulate your two sets of results in a table like this:

C R values	Time constant: calculated from formula (s)	Time constant: taken from graphs (s)
Set 1		
Set 2		
Set 3		

Conclusion

You should comment on the shape of the graphs: say what you think the graphs show. Are they what you would expect? For each combination of resistance and capacitance, estimate the time constant from the graph (the time constant should be the time taken to rise to 63% of the final voltage or fall by 37% of the initial voltage depending on whether the curve is for charge or discharge). Compare these values with the values that you calculated. If they are not the same, suggest possible reasons for the difference (such as component tolerance — the 1000 µF capacitor is unlikely to have a value of exactly 1000 µF).

ACTIVITY 7

1 Sketch and explain the shape of the magnetic field produced by the component shown in figure 1 and justify this in relation to the principles of electromagnetism and magnetic circuits. Indicate the position and polarity of the North and South magnetic poles.

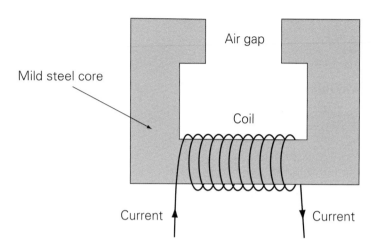

Figure 1 A magnetic component

You need to apply the right-hand rule to determine the direction of the magnetic field produced by the coil. Remember that when using the right-hand rule, the right thumb points in the direction of the current and the fingers (in the grasp position) indicate the direction of the magnetic flux, as shown in figure 2. You also need to remember that the arrows on the flux lines will indicate the direction of the movement of a free North pole. You should be able to use this information to sketch the magnetic field lines and reason out which side of the air gap will form the North magnetic pole and which will form the South magnetic pole.

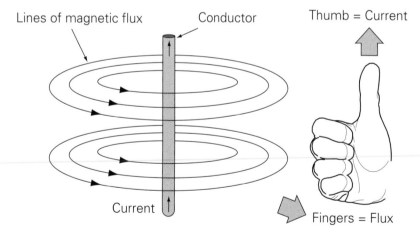

Figure 2 The right-hand rule allows you to determine the direction of the magnetic field

2 The data shown in table 1 refers to the mild steel used for the core of the material shown in figure 1. Plot the *B–H* curve for the material and identify the onset of saturation.

Table 1 B–H data

Field strength, H (A/m)	250	500	750	1000	1500	2000	3000	4000	5000	6000
Flux density, B (T)	0.85	1.1	1.27	1.35	1.45	1.5	1.57	1.63	1.67	1.7

You need to plot a graph showing flux density (*B*) against magnetic field intensity (*H*). This graph should be similar to the one shown in figure 3. You can then use the graph to identify the onset of saturation. At this point, there is no further significant increase in flux density for any further increase in magnetic field intensity.

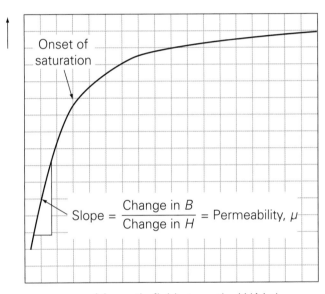

Magnetic field strength, *H* (A/m) ⟶

Figure 3 A B–H curve showing the onset of saturation

3 Calculate the permeability of the mild steel core in question 1 when *H* = 500 A/m.

You need to construct a tangent to the graph at the value of *H* that you were given in the question (500 A/m). The slope of the graph at this point (change in *B* divided by a corresponding change in *H*) will allow you to determine the permeability of the material. Figure 3 shows you how this is done.

ACTIVITY 8

Answer the following questions on inductors.

a) Explain what is meant by the self-inductance of a coil.

b) The field coil of a small motor comprises of 90 turns of enameled copper wire wound on a steel core having a relative permeability of 2000, a mean length of 0.05 m and a cross-sectional area of 8×10^{-4} m². Calculate the current required to produce a flux of 0.5 mWb in the field coil.

To answer (a) you need to explain that when current flows in a coil it gives rise to a magnetic flux that links with the coil. You will you also need to say that whenever the current in the coil changes there will be a corresponding change in the flux linkage and an emf will be induced to in the coil. You should continue by saying that because this emf is caused by changes in the coil itself, it is therefore referred to as a self-induced emf.

To answer (b) you should start by calculating the flux density, B. To do this you should use:

$$B = \frac{\Phi}{A}$$

where Φ is the flux and A is the cross-sectional area.

Next you can find the field strength using:

$$H = \frac{B}{\mu_0 \mu_r}$$

where B is the flux density that you have just calculated, μ_0 is the absolute permeability and μ_r is the relative permeability of the steel core.

Finally, you should be able to find the field current using the values that you have just calculated. To do this you should use:

$$I = \frac{Hl}{H}$$

where l is the mean length of the magnetic path and N is the number of turns.

ACTIVITY 9

Create an A2 size poster that describes three applications of electromagnetism and induction. Your poster should use simple language and should be designed to be easily understood by someone with limited technical knowledge. You should include illustrations of each of your applications using line drawings or photographs and a small amount of explanatory text.

Your poster could be based on any three of the following electromagnetic applications:

- inductors
- transformers

- motors
- generators
- relays
- solenoids.

Explain each of your chosen applications using a simple labelled sketch showing each of the component parts (such as coil, iron or steel core, supply connections, etc). Show how the magnetic field is produced and indicate the shape of the field using lines of flux drawn on the diagram. You might also find it useful to add a colour photograph taken using a digital camera, grabbed from the web, or scanned from a catalogue or datasheet.

You should include a few key points that relate to the fundamental principles that underpin each of the applications that you have chosen. For example, in the case of a transformer you might want to mention:

- the magnetic flux created by current flowing in the primary winding links with the secondary winding
- that the ratio of primary turns to secondary turns determines whether the transformer is a step-up or step-down component
- that the alternating flux coupled into the secondary winding generates an induced secondary voltage.

Finally, say where each of the applications is used. For example, in the case of a transformer:

- isolating one circuit from another (safety)
- stepping-up or stepping-down voltages in a power supply
- matching a microphone to the input of an amplifier
- matching a loudspeaker to the output of an amplifier.

Transformers

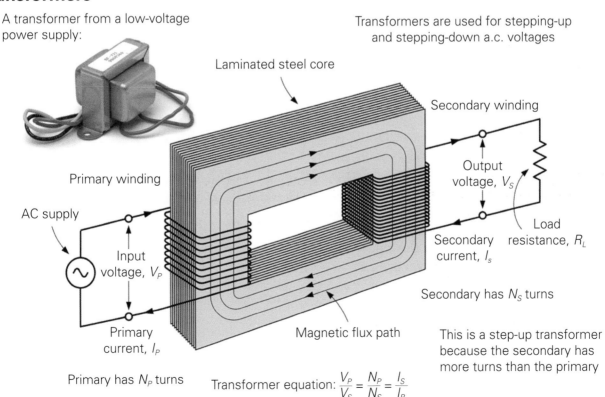

A transformer from a low-voltage power supply:

Transformers are used for stepping-up and stepping-down a.c. voltages

Laminated steel core

Secondary winding

Primary winding

AC supply

Input voltage, V_P

Primary current, I_P

Primary has N_P turns

Magnetic flux path

Output voltage, V_S

Load resistance, R_L

Secondary current, I_s

Secondary has N_S turns

This is a step-up transformer because the secondary has more turns than the primary

Transformer equation: $\dfrac{V_P}{V_S} = \dfrac{N_P}{N_S} = \dfrac{I_S}{I_P}$

93

ACTIVITY 10

Investigate the input and output waveforms and voltages for the AC circuit C–R network shown in figure 1 using an oscilloscope and a multimeter. Your investigation should be based on the following:

a) Display at least two complete cycles of the input and output voltages using the oscilloscope and use this to make a sketch of the input and output waveforms to a common time scale. Include labelled axes of voltage and time on your sketch.

b) Measure the peak-to-peak voltage of the input and output waveforms using the display on the oscilloscope.

c) Measure the value phase angle between the input and output waveforms using the display on the oscilloscope.

d) Use the multimeter to measure the rms voltage at the input and output.

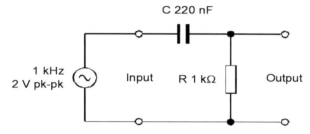

Figure 1 C–R circuit network

94

Note that the multimeter will read the rms voltage whilst the waveforms displayed on the oscilloscope will be easier to measure if you use the peak-to-peak value. Your investigation of the circuit will involve measuring the phase shift between the two waveforms.

You will need to connect the oscilloscope and the multimeter as shown in figure 2.

You will need to trigger the time base of the oscilloscope using the input voltage that you have connected for display using the Y1 channel. Figure 3 shows the initial settings for a typical oscilloscope (note that, for clarity, we have just shown the Y1 input connection on this diagram). When the oscilloscope has been correctly adjusted you will be able to measure the phase difference between the input and output waveforms, as shown in figure 4. Note that one complete cycle refers to 360°.

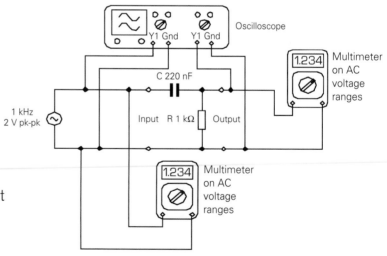

Figure 2 AC C–R circuit network

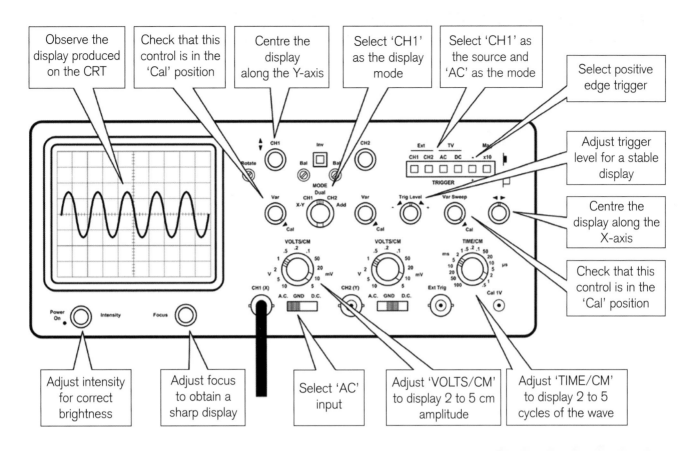

Figure 3 Oscilloscope adjustments (Y1 channel only shown)

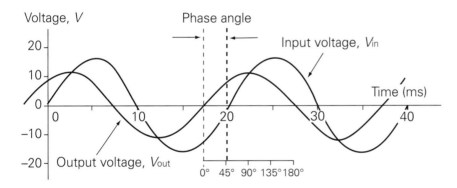

Figure 4 Using a waveform display to measure phase difference

ACTIVITY 11

Investigate the operation of simple DC power supplies that use half-wave and full-wave rectification, as shown in figures 1 and 2, respectively. Your investigation should include the following:

a) Measurement of the DC output voltage for different values of load resistance and reservoir capacitance.

b) Measurement of the ripple present on the DC output voltage for different values of load resistance and reservoir capacitance.

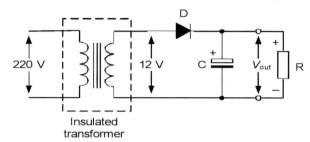

Figure 1 Simple DC power supply using half-wave rectification

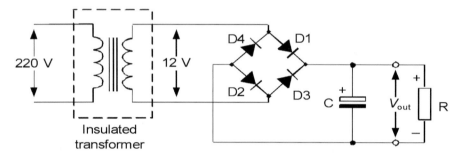

Figure 2 Simple DC power supply using full-wave bridge rectification

For both types of power supply, you will need to connect the oscilloscope and the multimeter as shown in figure 3. The multimeter should be set to the DC 20 V range and the oscilloscope should be adjusted so that the ripple present on the output voltage can be clearly seen.

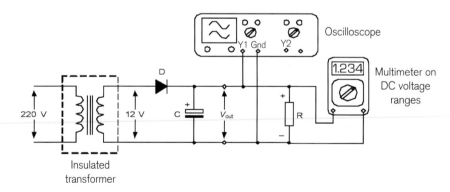

Figure 3 Connections for the oscilloscope and multimeter

Measure and record the output voltage and sketch the output waveform for a variety of different component values in both circuit arrangements, as shown in tables 1(a) and 1(b) on the next page. You can measure the ripple voltage as shown in figure 4. Note that in a perfect power supply the ripple voltage would be zero but in practice it may be as much as several hundred millivolts.

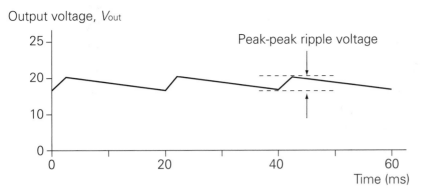

Figure 4 Using the displayed waveform to measure the ripple voltage

Table 1(a) Measurements made for the half-wave rectifier power supply

	C = not used R = 100 Ω 3 W	C = 100 µF R = 100 Ω 3 W	C = 100 µF R = 220 Ω 2 W	C = 220 µF R = 220 Ω 2 W	C = 470 µF R = 220 Ω 2 W
DC output voltage (from multimeter) (V)					
Ripple voltage (from oscilloscope) (V pk-pk)					

Table 1(b) Measurements made for the full-wave rectifier power supply

	C = not used R = 100 Ω 3 W	C = 100 µF R = 100 Ω 3 W	C = 100 µF R = 220 Ω 2 W	C = 220 µF R = 220 Ω 2 W	C = 470 µF R = 220 Ω 2 W
DC output voltage (from multimeter) (V)					
Ripple voltage (from oscilloscope) (V pk-pk)					

Conclusion

You should comment on the readings that you obtained. Are they what you would expect? Which of the circuits was best? Why was this? What effect does (a) the reservoir capacitor value and (b) the load resistor value have on the operation of the circuit? If you were designing a practical power supply, what values would you recommend?

This section focuses on grading criteria P1, P2, P3, P4, P5, P6, P7, P8, P9; M1, M2, M3, M4 and aspects of D1 and D2 from Unit 6 – Mechanical Principles and Applications.

Learning outcomes

1 Be able to determine the effects of loading in static engineering systems
2 Be able to determine work, power and energy transfer in dynamic engineering systems
3 Be able to determine the parameters of fluid systems
4 Be able to determine the effects of energy transfer in thermodynamic systems

Content

1) Be able to determine the effects of loading in static engineering systems

Non-concurrent coplanar force systems: graphical representation, eg space and free body diagrams; resolution of forces in perpendicular directions, eg $F_x = F\cos\theta$, $F_y = F\sin\theta$, vector addition of forces, conditions for static equilibrium ($\Sigma F_x = 0$), ($\Sigma F_y = 0$), ($\Sigma M = 0$), resultant, equilibrant, line of action.

Simply supported beams: conditions for static equilibrium; loading (concentrated loads, uniformly distributed loads, support reactions).

Loaded components: elastic constants (modulus of elasticity, shear modulus); loading (uniaxial loading, shear loading); effects, eg direct stress and strain including dimensional change, shear stress and strain including dimensional change, shear stress and strain, factor of safety.

2) Be able to determine work, power and energy transfer in dynamic engineering systems

Kinetic principles: equations for linear motion with uniform acceleration ($v = u + at$, $s = ut + \frac{1}{2}at^2$, $v^2 = u^2 + 2as$, $s = \frac{1}{2}(u + v)t$.

Kinetic parameters: eg displacement (s), initial velocity (u), final velocity (v), uniform acceleration (a).

Dynamic principles: D'Alembert, principle of conservation of momentum, principle of conservation of energy.

Dynamic parameters: eg tractive effort, braking force, inertia, frictional resistance, gravity, mechanical work ($W = Fs$), power dissipation (average power = W/t, instantaneous power = Fv), gravitational potential energy (PE = mgh), kinetic energy (KE = $\frac{1}{2}mv^2$).

3) Be able to determine the parameters of fluid systems

Thrust on a submerged surface: hydrostatic pressure, hydrostatic thrust on an immersed plane surface ($F = \rho gAh$); centre of pressure of a rectangular retaining surface with one edge in the free surface of a liquid.

Immersed bodies: Archimedes' principle; fluid, eg liquid, gas; immersion of a body, eg fully immersed, partly immersed, determination of density using floatation and specific gravity bottle.

Design characteristics of a gradually tapering pipe: eg volume flow rate, mass flow rate, input and output flow velocities, input and output diameters, continuity of volume and mass for incompressible fluid flow.

4) Be able to determine the effects of energy transfer in thermodynamic systems

Heat transfer: heat transfer parameters, eg temperature, pressure, mass, linear dimensions, time, specific heat capacity, specific latent heat of fusion, specific latent heat of vaporisation, thermal efficiency, power rating, linear expansivity; heat transfer principles, eg sensible and latent heat transfer, linear expansion; phase, eg solid, liquid, gas.

Thermodynamic process equations: Boyle's law ($pV = Constant$), Charles' law ($V/T = Constant$), general gas equation ($pV/T = Constant$), characteristic gas equation ($pV = mRT$); process parameters, eg absolute temperature, absolute pressure, volume, mass, density.

Grading criteria

P1 calculate the magnitude, direction and position of the line of action of the resultant and equilibrant of a non-concurrent coplanar force system containing a minimum of four forces acting in different directions

You need to know that for our purposes a non-concurrent coplanar force system is several forces that are all in the same plane but acting in different directions on a structural component. 'Non-concurrent' means that the lines of action of the forces don't pass through a single point. If they did, the forces would be 'concurrent' and you should already be able to determine the resultant and equilibrant of these simple force systems, either graphically or by calculation. To achieve P1, you will need to calculate the magnitude, direction and sense of the resultant and equilibrant for a non-concurrent force system and also the position of their line of action measured from some given point in the system. You will need to resolve the forces into perpendicular components in the x and y directions and calculate their moments about the given point. You will need to adopt a suitable sign convention for the directions of the force components and their moments. You must then total up the two sets of components and apply Pythagoras' theorem to calculate the magnitude of the resultant and equilibrant. The direction of their line of action can be found by applying simple trigonometry. Finally, you must find the perpendicular distance of their line of action from the given point, which can be done by equating the moment of the resultant to the sum of the separate moments.

P2 calculate the support reactions of a simply supported beam carrying at least two concentrated loads and a uniformly distributed load

A simply supported beam is one with two supports. In practice the supports are often in the form of rollers that allow the beam to flex and move freely when loaded or when there is a temperature change. Sometimes a beam may overhang its supports at one or both ends. Concentrated loads act at a point whilst uniformly distributed loads are spread out evenly along a beam. To achieve this

assessment criterion, you will need to know about the conditions for static equilibrium of a body and apply them to find the load that each support is carrying. One condition is that the sum of the moments taken about any point on the beam must be zero. Clockwise moments are usually taken to be positive and anticlockwise to be negative. By using this sign convention and taking moments about each of the support points in turn, you will be able to calculate the support reactions. When calculating the moment of a uniformly distributed load, you should consider it to act at its centroid, ie the mid-point of its span.

P3 calculate the induced stress, strain and dimensional change in a component subjected to direct uniaxial loading and the shear stress and strain in a component subjected to shear loading

You need to know that uniaxial forces are equal and opposite along the same line of action. They may be tensile or compressive. Shearing forces are also equal and opposite but don't act along the same line. They have parallel lines of action. To achieve this assessment criterion, you will need to calculate the direct stress in a given component and then use it, along with the given value of the modulus of elasticity of the material, to calculate the direct strain. You can then use this to calculate the dimensional change that takes place due to the loading. You will also need to calculate the shear stress and shear strain that is induced in a given component. The shear strain can be calculated from the shear modulus of the material, if it is known, or from the dimensional change that occurs if this has been measured.

P4 solve three problems that require the application of kinetic and dynamic principles to determine unknown system parameters

The three problems that you need to solve must involve the use of the three dynamic principles listed above. The first could be a problem involving the use of D'Alembert's principle to determine the force needed to accelerate an object and then to calculate the work done and power required. This is really the application of Sir Isaac Newton's three laws of motion. When we use D'Alembert's principle, we treat a moving body as if it were frozen in time and draw a free body diagram showing the forces that are acting on it. For a vehicle being accelerated up a hill it would appear as follows.

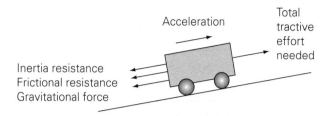

This is really the application of Newton's 2nd and 3rd laws of motion. The inertial resistance is calculated using the well-known formula $F = ma$, which is derived from Newton's 2nd law of motion. The 3rd law says that to every action there is an equal and opposite reaction. The total force needed to accelerate the body is the active force. It is equal and opposite to the three kinds of resistance, which are reactive forces.

The second problem could involve the use of the principle of conservation of momentum. It can be used to predict the final velocity when two bodies that are travelling at different velocities collide and become locked together. Alternatively it can be used to predict the final velocities of separation when an object such as a space vehicle separates by means of a controlled explosion into two separate sections. When bodies collide, there is a loss of kinetic energy from the system which is given off in the form of heat and sound waves. When bodies separate, as with a space vehicle, there is an increase in the kinetic energy of the system that is received from the explosion.

The third problem could also involve the use of the principle of conservation of energy. You can apply it to the same kind of dynamic system problems as those solved using Newton's laws and D'Alembert's principle. It is an alternative method, and sometimes a little quicker depending on the information that you are given. Instead of finding the force required to overcome the total resistance of a body, it enables you to find the total work done from which you can then find the force and the power required.

P5 calculate the resultant thrust and overturning moment on a vertical rectangular retaining surface with one edge in the free surface of a liquid

To achieve this assessment criterion, you first need to be able to calculate the thrust of a liquid on a rectangular retaining surface such as the side of a tank, a wall or a lock gate. The formula for this is given above, ie $F = \rho gAh$, where ρ is the liquid density, g is the acceleration due to gravity, A is the rectangular wetted area of the retaining surface and h is the depth of the centroid of the wetted area, which will be at *half* of the depth contained. You then need to know the point at which this thrust is considered to act on a rectangular retaining surface. This is called the centre of pressure. It is always below the centroid and you can prove using calculus that it is at two-thirds of the depth below the free surface. With this, and the value that you have calculated for the thrust, you can calculate the overturning moment about the base of a wall or gate that is exerted by the liquid. Remember that a turning moment is the thrust multiplied by the perpendicular distance from the turning point at the base of the surface, ie one-third of the wetted depth.

P6 Determine the up-thrust on an immersed body

To achieve this assessment criterion, you need to apply Archimedes' principle. This says that the up-thrust on a body is equal to the weight of the fluid displaced. In other words, for a body fully immersed in a liquid, the up-thrust is equal to the weight of the liquid that has the same volume as the body. Provided that you know the volume of an immersed body, or the volume below the free surface for a floating body, you can use this together with the specific weight of the liquid to calculate the up-thrust.

P7 use the continuity of volume and mass flow for an incompressible fluid to determine the design characteristics of a gradually tapering pipe

We generally say that liquids cannot be compressed. In fact, they can be compressed a little at very high pressures. We also assume that the flow through a pipe is steady and that the pipes are running full. With steady flow, the flow velocity at any cross-section of a pipe is steady. When this is the case, the volume and mass per second of the liquid entering a pipe must be the same as that leaving it, even if the diameter of the pipe has changed. This is what we mean by *continuity of volume and continuity of mass*. The continuity equations relate input and output conditions and you will need to use them to achieve this assessment criterion. Given the volume or mass flow rates, you should be able to calculate the flow velocity at any given cross-section. Alternatively, if you know the flow velocity at a given cross-section, you should be able to calculate the volume and mass flow rates.

P8 calculate dimensional change when a solid material undergoes a change in temperature and the heat transfer that accompanies a change in temperature and phase

To achieve the P8 assessment criterion, you need to be able to calculate the change in length of a component that takes place when its temperature changes. You will need to know about the *linear expansivity* of a material which is also sometimes called the *coefficient of linear expansion*. To calculate the change in length of a component you must multiply the linear expansivity by the original length (measured in metres or millimetres – it doesn't matter which) and the temperature change. The answer will be in the same units that you used for the original length.

Also to achieve the P8 assessment criterion, you will also need to calculate the *sensible* and *latent heat* transfer that might accompany a temperature change. To calculate the sensible heat transfer, you must know the *specific heat capacity* of a material or substance. You multiply this by the mass of material and its temperature rise and the answer will be in Joules. Sometimes a substance will melt or evaporate when it receives heat energy. We then say that a change of phase has occurred and this always takes place at a constant temperature even though heat transfer is still taking place. You probably know that this is called *latent heat* transfer. To find the latent heat transfer that occurs during a phase change you need to know the *specific latent heat of fusion* or the *specific latent heat of vaporisation*, depending on whether the substance is changing from solid to liquid or from liquid to gas. Their units are Joules per kilogram and all you need to do is multiply by the mass of the substance to find the latent heat transfer measured in Joules.

P9 solve two or more problems that require application of thermodynamic process equations for a perfect gas to determine unknown parameters of the problems

A perfect or 'ideal' gas is one that obeys the gas laws (Boyle's and Charles') and the formulae that we derive from them, at all temperatures and pressures. Real gases such as hydrogen, oxygen and nitrogen are not quite ideal but they obey the gas laws fairly closely and we often assume that they behave like a perfect gas. To achieve this assessment criterion, you will need to solve two or more problems on the expansion and compression of gases, making use of the thermodynamic process equations listed in the Content section on p.100. When you are solving these problems, remember that temperatures must be absolute temperatures, measured in Kelvin. If you are given a temperature in degrees Celsius, you must add on 273 degrees to change it to Kelvin. Also, your pressures must be absolute pressures. If you are given the reading from a pressure gauge, you must add on atmospheric pressure, and on average its value is 101.325 kPa. If you don't use absolute temperature and pressure, your answers will not be correct.

M1 calculate the factor of safety in operation for a component subjected to combined direct and shear loading against given failure criteria

You need to know that a structural component can be considered to have failed when fracture occurs and sometimes also before fracture when the stress has exceeded its elastic limit value. You will be told which of these failure criteria to use in your calculations. To achieve this assessment criterion, you will need to calculate the direct stress and the shear stress in a component that is subjected to both kinds of loading at the same time. You will then need to compare these with the given values of direct stress and shear stress at which failure is considered to have occurred to calculate two possible values of factor of safety. The lower value is the factor of safety in operation.

M2 determine the retarding force on a freely falling body when it impacts upon a stationary object and is brought to rest without rebound, in a given distance

To achieve M2, you will need to solve a problem such as looking at a pile-driving hammer forcing a pile into the ground. Alternatively, the resistance of metal to a drop hammer that is being used to forge the metal into a shape. Problems such as this can be solved by applying the equations for linear motion or application of the principle of conservation of momentum.

M3 determine the thermal efficiency of a heat transfer process from given values of flow rate, temperature change and input power

The heat transfer processes that you will be considering are the sort that take place in water heaters that you find in central heating systems and process plants; also, in air heaters and coolers that you might find in air conditioning systems. There is a steady flow of liquid or gas through these systems, during which a temperature change occurs. Energy is supplied to the system but not all of it goes to producing the temperature change. The thermal efficiency of a system is a measure of its effectiveness, and to calculate its value you must divide the energy received per second by the energy supplied per second. This is the same as saying input power divided by output power because power, measured in Watts, is energy transfer per second. Your answer will always be a decimal fraction and it is standard practice to multiply by 100 to change it to a percentage.

M4 determine the force induced in a rigidly held component that undergoes a change in temperature

To achieve the M4 assessment criterion, you need to be able to calculate the change in length of a component that takes place when its temperature changes (as for P8). But you need to go on to understand that when a rigidly held component is heated or cooled it is prevented from expanding or contracting. We say that *thermal strain* is present which produces *thermal stress* in the material. You will recall that strain is change in length divided by original length. Here the length doesn't actually change but in its place we use the value of the expansion or contraction that has been prevented from taking place. To achieve the M4 assessment criterion, you will need to use the value of thermal strain and the modulus of elasticity of the material to calculate thermal

stress. You can then multiply this by the cross-sectional area of the component to find the tensile or compressive force in the material.

D1 compare and contrast the use of D'Alembert's principle with the principle of conservation of energy to solve an engineering problem

To achieve D1, you will again need to look at a problem such as a pile-drive working on a building site or a drop hammer forging metal into a shape. You will need to solve the problem given in two ways – applying the equations for linear motion and the application of the principle of conservation of momentum – and then comparing your results. Note the use of 'contrast' in this criterion: you can go on to use the two results as an answer check and should say which method is best suited to solving a particular problem, given the different information that you may have in each case.

D2 evaluate the methods that might be used to determine the density of a solid material and the density of a liquid

The methods that you should consider are a) application of Archimedes' principle, b) use of a specific gravity bottle. For both methods you need an accurate cross-beam type of chemical balance and a set of precision weights. You must be able to set up and operate the balance correctly and handle the weights in the approved manner. You also need to know the density of water which is of course $1000 \, \text{kg m}^{-3}$. The immersion method which makes use of Archimedes' principle and can be used to find the density of both solids and liquids. For a solid, it involves accurately weighing a specimen when suspended in air and then when submerged in water. The specimen must of course be denser than water or it would float, and this is a limitation of the method. To find the density of an unknown liquid, the solid specimen is weighed a third time when submerged in the liquid. The data can then be used to calculate densities.

The specific gravity bottle is a small rounded flask made of clear glass with an accurately ground stopper. The term *specific gravity* has been replaced by the modern name, *relative density*. The bottle holds precisely 50 cc of liquid with the stopper in position. The stopper has a small hole drilled through it, which allows excess liquid to escape. The bottle must be accurately weighed when empty, when full of water and then when full of a second liquid whose density it is required to find. To determine the density of a solid material, the bottle is again weighed when empty and when full of water. A sample of material small enough to fit in the bottle is then weighed and placed in the bottle. This is then filled with water, the stopper is replaced and the bottle weighed

again. The data can then be used to calculate densities.

As you will see, the procedures are an exercise in accurate weighing and to achieve this assessment criterion you should ideally be given the opportunity to use both methods. You will then be able to compare the results and contrast the two methods.

ACTIVITY 1

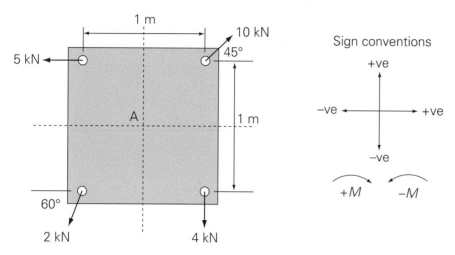

Calculate the magnitude and direction of the resultant and equilibrant of the forces acting on the plate shown above and the perpendicular distance of their line of action from the point A, at the centre of the plate.

The above drawing is called a *space diagram*. It shows the plate with the external forces acting on it in isolation from the structure to which it is attached. You will need to resolve the forces into their horizontal and vertical components. It is best that forces that are not already horizontal or vertical should have their angles measured to the horizontal. Their horizontal (x-direction) and vertical (y-direction) components will then be $F_x = F\cos\theta$ and $F_y = F\sin\theta$, respectively. You should use the sign convention that horizontal components acting to the right are positive and those to the left are negative. Also, upward vertical components are positive and downward ones are negative. It is important that the components carry the correct positive or negative sign and it is good practice to list your results in the form of a table: note the symbol ΣF in the table means that this is the sum of the forces and ΣM is the sum of the moments.

Force	F_x (Horz. components)	F_y (Vert. components)	Moment of F_x about the point A	Moment of F_y about the point A
10 kN				
4 kN				
2 kN				
5 kN				
Totals	$\Sigma F_x =$	$\Sigma F_y =$	$\Sigma M =$	

The next step is to calculate the moments of the components about the point A, at the centre of the plate. Here you disregard the sign convention for the forces and adopt the one for moments. That is, that clockwise moments are positive and anticlockwise moments are negative. These should also be listed in the table and you now have to total up the horizontal components, then the negative components and then the moments.

You should find that the totals ΣF_x and ΣF_y are both positive and can be shown on a simple force vector diagram as follows.

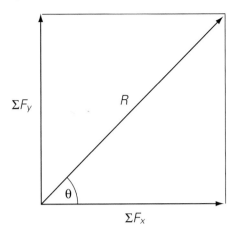

The final steps are to calculate the value of the resultant and equilibrant using Pythagoras' theorem.

ie $\quad R^2 = (\Sigma F_x)^2 + (\Sigma F_y)^2$

The equilibrant will of course be equal and opposite to the resultant and you can find the angle θ of their line of action by simple trigonometry.

ie $\quad \tan\theta = \Sigma F_y / \Sigma F_x$

Finally, you can find the perpendicular distance a, of their line of action from the point A, at the centre of the plate by equating the moment of the resultant R about A, to the sum of the moments in your table.

ie $\quad Ra = \Sigma M$

You should find that the value of ΣM in your table is positive, which means that the resultant must have a clockwise turning moment about A. The position of the line of action of the resultant must thus be as follows.

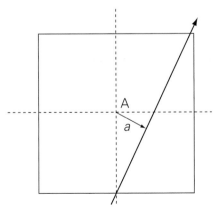

ACTIVITY 2

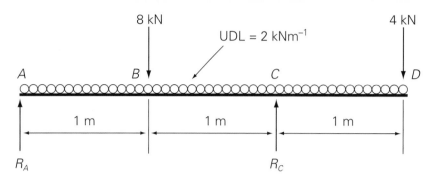

You are required to find the support reactions R_A and R_C for the above simply supported beam. The beam has two concentrated loads and a uniformly distributed load (UDL) of 2 kN per metre length.

The downward concentrated loads and the uniformly distributed load are due to the force of gravity and are said to be *active loads*. The loads carried by the supports are said to be *reactive loads* which is why they are called *support reactions*. There are three conditions that apply for this beam to be in a state of static equilibrium.

1) The algebraic sum of the horizontal forces must be zero, ie $\Sigma F_H = 0$. In this case there are no horizontal forces so there is no need to consider them.

2) The algebraic sum of the vertical forces must be zero, ie $\Sigma F_x = 0$. If we use the same sign convention as in Activity 1, this states that the downward negative forces must be cancelled out by the two upward positive forces.

3) The algebraic sum of the moments of the forces taken about any point along the beam must be zero, ie $\Sigma M = 0$. If we again use the same sign convention as in Activity 1, this states that the positive clockwise moments about any point on the beam must be cancelled out by the negative anticlockwise moments.

You should begin be taking moments about the support at A to find the reaction R_C of the support at C. The full uniformly distributed load can be considered to act at its centroid, and when you have calculated its total value, it can be treated as just another concentrated load. You should then take moments about the support at C to find the reaction R_A of the support at A. This will be a little more complicated because the UDL acts on either side of C and you will need to consider it in two separate parts, each acting as a concentrated load at its centroid. One will be at the mid-point of the span AC, giving an additional load at B. The other will be at the mid-point of the overhang, CD.

When you have calculated the values of R_A and R_C you should check your calculation by applying the second of the above conditions for static equilibrium, ie $\Sigma M = 0$. The sum of R_A and R_C that is upward and positive should be cancelled out by the total downward load, which is negative.

ACTIVITY 3

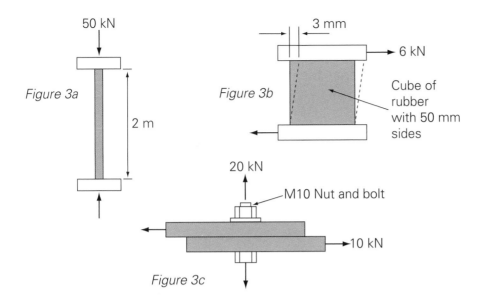

Figure 3a

2 m

Figure 3b

Cube of rubber with 50 mm sides

20 kN

M10 Nut and bolt

10 kN

Figure 3c

Task 1

Figure 3a shows a tubular steel strut of outer diameter 40 mm and inner diameter 30 mm which carries a compressive load of 50 kN. The strut is 2 m long and the modulus of elasticity of the material is 200 GPa. Calculate the induced compressive stress, the compressive strain and the change in length that takes place during loading.

Task 2

Figure 3b shows a mounting block which consist of a 50 mm cube of rubber bonded between steel plates. The shear modulus of the rubber is 120 MPa. The block is subjected to a shearing force of 6 kN which causes the plates to move 3 mm over each other as shown. You are required to calculate the shear stress and the shear strain in the rubber.

Task 3

Figure 3c shows two plates that are held together by a 10 mm diameter nut and bolt. The bolt is tensioned to 20 kN and subjected to a shearing force of 10 kN from the plates. The ultimate tensile strength of the bolt material is 500 MPa and its shear strength is 300 MPa. You are required to calculate the factor of safety in operation.

For each of the above tasks you should set out your calculations in a logical fashion, stating clearly what you are about to do before each calculation. Be careful always to state the units of your answers and make full use of the EXP and ENG functions on your electronic calculator when you are entering and recording multiples and sub-multiples.

ACTIVITY 4

Solve the following three problems on dynamic systems:

Task 1

A vehicle of mass 850 kg accelerates uniformly from rest to a velocity of 70 km h^{-1} in a time of 12 seconds while ascending a 1 in 6 gradient (sine). The frictional resistance to motion is 0.6 kN. Making use of D'Alembert's principle, determine:

i)   the tractive effort between the wheels and the road surface.

ii) the work done in ascending the slope

iii) the average power developed by the engine

iv) the power developed by the engine as the vehicle reaches its maximum velocity.

In this first problem you will need to convert the final velocity to m s^{-1} and use the equations for linear motion to calculate the distance travelled and the acceleration of the vehicle. You can then use the acceleration to calculate the inertial resistance. You will also need to find the resistance due to gravity. This is the component of the vehicle's weight that acts parallel to the slope. The frictional resistance is given, so you can then draw the free body diagram and calculate the total tractive effort required. Finally, the work done and the power required from the engine can be calculated using the formulae given in the Content section on p. 98.

Task 2

A space vehicle travelling at a velocity of 5000 km h^{-1} separates by a controlled explosion into two sections of mass 900 kg and 350 kg. The two parts carry on in the same direction with the heavier section moving at 2500 km h^{-1}. Making use of the principle of conservation of momentum, determine the velocity of the lighter section. Determine also the change in the total kinetic energy of the system.

In this second problem you should first convert the given velocities to m s^{-1}. You will then need to equate the initial and final momentum of the system and transpose it to make the velocity of the lighter section the subject. When you have done this, you will need to calculate the initial and final kinetic energy of the system. You should find that there has been an increase as some of the energy from the controlled explosion is converted into kinetic energy.

Task 3

A lift cage of mass 550 kg accelerates upwards from rest to a velocity of 5 m s^{-1} whilst travelling through a distance of 12 m. Frictional resistance to motion is 250 N. Making use of the principle of conservation of energy, determine:

i)   the work input from the driving motor

ii) the tension in the lifting cable

iii) the maximum power developed by the driving motor.

In this third problem, application of the principle of conservation of energy will enable you to directly calculate the work input from the driving motor. It is the sum of the gain in potential energy, the gain in kinetic energy and the work done in overcoming friction. You can then use it to calculate the tension in the lifting cable and the maximum power developed using formulae given in the Content section on p. 98.

As for the previous activities, you should set out your calculations in a logical fashion, stating clearly what you are about to do before each calculation. Always state the units of your answers and make full use of the ENG function on your electronic calculator when you are recording multiples and sub-multiples.

ACTIVITY 5

A pile-driver hammer of mass 150 kg falls freely through a distance of 5 m to strike a pile of mass 400 kg and drives it 75 mm into the ground. The hammer does not rebound when driving the pile. Determine the average resistance of the ground. You are required to solve this problem in two ways: a) by making use of the principle of conservation of momentum and D'Alembert's principle; b) by making use of the principle of conservation of energy.

With the first method you will need to calculate the velocity of the hammer immediately before it strikes the pile using an appropriate equation of motion. You will then need to apply the principle of conservation of momentum to calculate the velocity of the pile and hammer immediately after the impact. The next step is to calculate the retardation of the pile and hammer as they are brought to rest. D'Alembert's principle can then be applied to find the resistance of the ground.

With the second method you will need to calculate the velocity of the hammer immediately before it strikes the pile using the principle of conservation of energy. As with the first method, you will then need to apply the principle of conservation of momentum to calculate the velocity of the pile and hammer immediately after the impact. You can then apply the principle of conservation of energy again to find the work done in bringing the pile and hammer to rest. You can then use this to find the average resistance of the ground. You should be able to conclude that both methods give the same answer but that for the information given, there may be a little less calculation required with the second method.

As with the previous activities, you should set out your calculations in a logical fashion, stating clearly what you are about to do before each calculation. Always clearly state the units of your answers and make full use of the ENG function on your electronic calculator when you are recording multiples and sub-multiples.

ACTIVITY 6

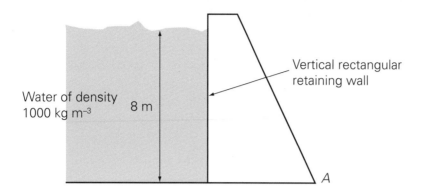

Water of density 1000 kg m⁻³

8 m

Vertical rectangular retaining wall

A

The water in a reservoir is held back by a vertical retaining wall as shown above. Find the thrust per metre length acting on the wall and the overturning moment per metre length acting about the point A.

Unless you are told otherwise, you can always assume that the density of the water is 1000 kg m⁻³. The other information that you require to calculate the thrust acting per metre length of the wall is the depth of the centroid of the wetted rectangular area, which is of course at half the wetted depth, and the acceleration due to gravity (9.81 m s⁻²). You can then use the formula that is given in the Content section on p. 98.

To find the overturning moment, you need to know the point at which the thrust can be considered to act, ie the position of the centre of pressure. This is at two-thirds of the depth of water or one-third of the distance up from the base of the wall. You can then use this and the thrust to calculate the overturning moment per metre length about the point A. In practice it is very unlikely that a retaining wall such as this would overturn. The weight of the wall material acting downwards though its centre of gravity produces a turning moment in the opposite direction which will be several times greater if the wall is properly designed.

A point worth remembering is that the centre of pressure is only at two-thirds of the depth of water if the retaining surface is rectangular. For other shapes, such as triangular or trapezoidal, etc, it is at a different depth.

ACTIVITY 7

Find the mass of a submarine from the following data.

The submarine floats completely submerged after taking 750 tonnes of water into its ballast tanks. The volume of the submarine is 1100 m³ and the density of seawater is 1020 kg m⁻³.

When the submarine is completely submerged and floating, the up-thrust is just equal to its weight plus the weight of water that it has taken on board. From Archimedes' principle you know that the up-thrust is equal to the weight of the volume of seawater that is equal to

110

the volume of the submarine. This enables you to calculate the value of the up-thrust, measured in newtons. From this you can find the mass of the submarine plus the water that it has taken on board. The mass of water can then be subtracted to give you the mass of the submarine itself.

To achieve the assessment criterion for this unit, you need to apply Archimedes' principle. This says that the up-thrust on a body is equal to the weight of the fluid displaced. In other words, for a body fully immersed in a liquid, the up-thrust is equal to the weight of the liquid that has the same volume as the body. Provided that you know the volume of an immersed body, or the volume below the free surface for a floating body, you can use this together with the specific weight of the liquid to calculate the up-thrust.

ACTIVITY 8

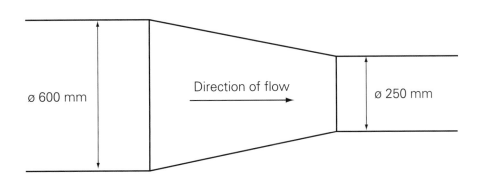

The pipe shown above is running full with oil of relative density 0.87 and the flow velocity at the entry to the tapered section is of 1.5 m s^{-1}. Find the flow velocity at the exit from the tapered section and mass flow rate of the oil. The density of water is 1000 kg m^{-3}.

The first step is to apply the continuity of volume equation which will enable you to calculate the flow velocity at the exit from the tapered section. You can also use the inlet velocity and diameter to find the volume flow rate measured in m^3 s^{-1}.

Before you can find the mass flow rate, you will need to find the density of the oil from its relative density and the density of water. You can then use it to convert the volume flow rate into mass flow rate measured in kg s^{-1}.

You might think about the way that the flow velocity changes with a change in pipe diameter. If the exit diameter had been 300 mm or 200 mm, how many times larger would the exit velocities be than the inlet velocity?

ACTIVITY 9

This is a practical activity in which you will gather data to find the density of a sample of metal and a sample of oil. You will be using the two methods mentioned previously, using the following equipment and procedures. Before you begin, you should carry out a little research to find out how the formulae used to calculate density using the immersion method, are derived from the application of Archimedes' principle. You should do the same for the specific gravity bottle so that you understand the purpose of the following procedures.

Equipment: Beam-type chemical balance and weights, bridge piece to place over one of the scale pans, length of fine cotton, 250 ml beaker, 50 cc specific gravity bottle, metal sample and oil sample.

Method 1: by immersion

1) Tie a length of cotton to the metal sample and attach to one end of the balance beam so that it is suspended just above the centre of the scale pan.

2) Carefully add weights to the other scale pan until balance occurs and record the mass of the metal sample.

3) Place the bridge piece across the scale pan under the suspended metal sample.

4) Fill the beaker with water and place it on the bridge piece so that the metal sample is fully immersed in it.

5) Again add weights to the other scale pan and record the mass reading with the metal submerged.

6) Empty the water from the beaker, dry it and fill it with the oil.

7) Repeat task 5 and record the mass reading with the metal submerged in the oil.

8) Calculate the density of the metal and oil samples from the data that you have gathered.

Recorded data

Mass of metal sample suspended in air, $m_1 =$
Mass recorded with metal sample immersed in water, $m_2 =$
Mass recorded with metal sample immersed in oil, $m_3 =$

Calculations

Density of metal sample, $\rho_M = m_1 \left[\dfrac{m_1}{m_1 - m_2} \right] \times \rho_W$

[where ρ_W is the density of water]

Density of oil sample, $\rho_O = \left[\dfrac{m_1 - m_2}{m_1 - m_3} \right] \times \rho_W$

Method 2: using a specific gravity bottle

1) Weigh the metal sample, if it is different to that used in Method 1, and record its mass.

2) Weigh the specific gravity bottle and stopper when empty and record the mass.

3) Weigh the specific gravity bottle when filled with water, with the stopper in position, and record the mass.

4) Place a sample of the metal in the water-filled specific gravity bottle, replace the stopper, weigh again and record the mass.

5) Empty and dry the specific gravity bottle, fill with the oil, replace the stopper, weigh again and record the mass.

6) Calculate the density of the oil and metal samples from the data that you have gathered.

Recorded data

Mass of metal sample, $m_1 =$

Mass of specific gravity bottle empty, $m_2 =$

Mass of specific gravity bottle filled with water, $m_3 =$

Mass of specific gravity bottle filled with oil, $m_4 =$

Mass of specific gravity bottle filled with water and metal sample, $m_5 =$

Calculations

Density of oil sample, $\rho_O = \dfrac{m_4}{m_3} \times \rho_W$

Density of metal sample, $\rho_M = \left[\dfrac{m_1}{m_1 - m_3 - m_5} \right] \times \rho_W$

Conclusions

Compare the values of the densities you obtained using the two methods.

Discuss any difficulties that you encountered and any possible sources of error.

Compare and contrast the advantages, disadvantages, limitations, etc, of the two methods.

ACTIVITY 10

A 10 mm diameter tie rod of effective length 200 mm and mass 0.15 kg is heated from 15°C to 150°C before being securely fixed between two rigid plates in an engineering assembly. The rod is then allowed to cool down to its original temperature so that it exerts a pulling force on the plates. You are required to find the heat energy received by the rod, the amount by which the tie rod expands and the force it exerts on the plates after cooling. The specific heat capacity of steel is 460 J kg^{-1} K^{-1}, its linear expansivity is 12×10^{-6} K^{-1} and its modulus of elasticity is 200 GPa.

Engineering components are often heated before assembly so that they exert a pulling or gripping force on each other after cooling down. Sometimes they are also cooled down to sub-zero temperatures before assembly so that they exert a pulling or gripping force after returning to room temperature.

When the rod is heated, you can calculate the heat energy received from its mass, specific heat capacity and temperature rise. You can also calculate the amount of expansion that occurs from its linear expansivity, temperature rise, and original length.

We must assume that the rod is quickly assembled in position and the next step is to divide the expansion by its original length to find the thermal strain that is set up after it has cooled down. We must also assume that the rod is elastic and that after cooling, the tensile stress induced is below the elastic limit for the material. You will recall that the modulus of elasticity is stress divided by strain, and you can now use this formula to find the tensile stress in the rod after it has cooled.

The next step is to calculate the cross-sectional area of the rod in m^2. Finally, you can multiply the tensile stress by the cross-sectional area to find the force exerted on the plates. As with previous activities, set out your calculations in a logical fashion, stating clearly what you are about to do before each calculation. The formulae are quite simple but you are dealing with multiples and sub-multiples and it is quite easy to make a mistake with units. You should make full use of the EXP function on your electronic calculator when you are entering the linear expansivity, modulus of elasticity and cross-sectional values. You should also make full use of the ENG function before writing down your answers.

ACTIVITY 11

A marine diesel engine cylinder has a diameter of 200 mm and the length of the piston stroke is 350 mm. When the piston is at bottom dead centre, air is taken in at a pressure of 100 kPa and temperature 25°C. At top dead centre the piston compresses the air into a clearance volume of 600 cc, at which point the pressure is 4.8 MPa. Find the temperature of the compressed air and its mass. The characteristic gas constant for air is 287 J kg^{-1} K^{-1}.

Diesel engines are known as compression ignition engines. They take in air and compress it into a very small volume. This causes the air temperature to rise so that when fuel is injected into the cylinders, it immediately starts to burn, producing a large increase in pressure. This forces the piston down the cylinder to drive the crankshaft and give output power. When the piston is at the start of its compression stroke, the crank is said to be at *bottom dead centre*, or *outer dead centre* position. When it is at the end of its compression stroke, it is said to be at *top dead centre*, or *inner dead centre* position.

Your first task in this activity will be to calculate the volume through which the piston moves. This is called the *swept volume*, and to find the initial volume of the air you must add on the given clearance volume. Even though the volumes are comparatively small, you should calculate them in m^3 and change the swept volume to m^3 before adding it on. You can now use the general gas equation to find the temperature of the air after compression. Remember that the temperature must be in Kelvin. We can assume, unless told otherwise, that the given pressures are absolute pressures, but change them to Pascals before putting them into the equation.

To find the mass of the air, you will need to use the characteristic gas equation. To calculate the mass in kg, the pressure must be in Pascals (Pa), the volume must be in m^3 and the temperature must be in Kelvin (K), otherwise the equation will not work. As with the other activities, you should make full use of the EXP function on your electronic calculator when you are entering the values of pressure and volume. You should also make full use of the ENG function before writing down your answers.

ACTIVITY 12

You are required to calculate the efficiency of an ice-making machine that takes in water at a temperature of 15°C and produces ice cubes at a temperature of −5°C. Water is taken in at the rate of 1 kg every 5 minutes and the input power to the machine is 300 W. The specific heat capacity of water is 4187 J kg^{-1}K^{-1}, the specific heat capacity of ice is 2 100 J kg^{-1}K^{-1} and the specific latent heat of fusion is 335 J kg^{-1}.

The thermal efficiency of the machine can be taken as the heat energy extracted per second divided by the power input. First, however, you need to calculate the mass flow rate in kgs^{-1}. There are then three steps to calculate the heat energy extracted from the water per second. You must calculate the heat energy extracted per second to cool the water down to 0°C. Then the latent heat extracted per second as the water at 0°C is changed to ice at 0°C. Then the heat energy extracted as the ice is cooled down to −5°C. These can then be added together, to give the total heat energy extracted per second. You should find that the latent heat extracted is small compared to the sensible heat. The units of your answers will be Watts. Finally, you can calculate the efficiency of the ice-maker. Your answer will be a decimal fraction which can be multiplied by 100 to give a percentage.

ANSWERS

ACTIVITY 1

1.71 kN, 51.3° to horizontal, 0.51 m from A

ACTIVITY 2

3.5 kN, 14.5 kN

ACTIVITY 3

Task 1

90.9 MPa, 455×10^{-6}, 0.909 mm

Task 2

2.4 MPa, 0.06

Task 3

1.96

ACTIVITY 4

Task 1

i)   3.37 kN
ii) 392 kJ
iii) 32.7 kW
iv) 65.3 kW

Task 2

3175 ms^{-1}, 0.775 GJ

Task 3

i)   74.6 kJ
ii) 6.22 kN
iii) 31.1 kW

ACTIVITY 5

26.7 kJ

ACTIVITY 6

314 kN, 837 kN m

ACTIVITY 7

372 tonnes

ACTIVITY 8

8.64 m s^{-1}, 369 kg s^{-1}

ACTIVITY 10

9.32 kJ, 0.324 mm, 25.4 kN

ACTIVITY 11

467°C, 0.0136 kg

ACTIVITY 12

81.8%

MARKED ASSIGNMENTS

UNIT 1 – BUSINESS SYSTEMS FOR TECHNICIANS

UNIT 6 – MECHANICAL PRINCIPLES AND APPLICATIONS

Unit 1 – Business Systems for Technicians
Sample assignment: Engineering companies

Unit:	1 – Business Systems for Technicians
Assignment:	1 – Engineering Companies
Covering:	P1, P2, P3, M1, D1
Learner:	Craig Evans
Assessor:	Andy Sharpe
Internal verifier:	Dan Mills
Date set:	12/11/07
Due date for completion:	15/02/08

This assignment comprises FIVE individual tasks. You must complete EACH of these tasks and submit your assignment in the form of a written report BEFORE the due date stated above. Please read the instructions carefully and make sure that you understand what is required before you start the assignment. For this assignment you are given three different engineering companies to base your answers on. These are Woodford Marine Diesels Ltd, Maybury Engineering plc and Jet Air Transport Services Ltd.

Unit 1: Assignment 1, Task 1 (P1)
Engineering sectors and functions

This task provides evidence for Unit 1, P1.

What you have to do

Identify the sectors that three given engineering companies do most business in and describe the engineering function that they carry out in that sector.

Evidence that you need to produce

You need to provide information about three separate engineering companies. The companies Woodford Marine Diesels Ltd, Maybury Engineering plc and Jet Air Transport Services Ltd are from three different sectors and may offer primary, secondary or tertiary engineering functions. You should include a description of the function they carry out in that sector.

Unit 1: Assignment 1, Task 2 (P2)
Company organisation

This task provides evidence for Unit 1, P2.

What you have to do

Describe the organisational types in terms of size, status and structure for the three given engineering companies.

Evidence that you need to produce

You should produce a brief written explanation together with an organisational chart showing the functions within each company and the links between them. Overall you need to comment about size, status and structure for the three given engineering companies. Your answer should link to the answer for Task 3 where you must describe the information flow within one of the organisations.

Unit 1: Assignment 1, Task 3 (P3)
Information flow

This task provides evidence for Unit 1, P3.

What you have to do

Explain information flow through an engineering company in relation to an engineering activity.

Evidence that you need to produce

You should provide a written explanation together with a diagram that shows how the functions within one of the companies are able to communicate effectively and how this supports business strategies in relation to an engineering activity. Overall you should comment on internal systems, the people involved, the types of information and the work ethics of communication. Your answer can be based on the work that you have done previously for Task 2 but your diagram should show the flow of information (including the type of information and its physical form) rather than the direct reporting lines.

Unit 1: Assignment 1, Task 4 (M1) Improvements to information flow

This task provides evidence for Unit 1, M1.

What you have to do

Identify and explain how improvements in information flow could enhance the functional activities of the engineering company.

Evidence that you need to produce

This is an extension of Task 3 and it should take the form of a brief written explanation together with any relevant charts or diagrams. You should also briefly explain the benefits of any changes that you recommend. Note that there is no need to repeat any of the work that you have previously submitted in relation to Task 3.

Unit 1: Assignment 1, Task 5 (D1) Evaluation of information flow

This task provides evidence for Unit 1, D1.

What you have to do

Evaluate the information flow through a company in relation to an engineering activity and explain how improvements in information flow could enhance the functional activities of the company.

Evidence that you need to produce

This is an extension of Task 4. You need to apply evaluative skills to explain how the information flow in Task 4 could enhance and improve the functional activities of the engineering company. Once again you should present your work in the form of a brief written explanation together with relevant charts and diagrams. Note that there is no need to repeat any of the work that you have previously submitted in relation to Task 4.

PASS LEVEL ANSWERS (ATTEMPTED)

Unit:	**1 Business Systems for Technicians**
Assignment:	**1 Engineering Companies**
Covering:	**P1, P2, P3, M1, D1**
Learner:	**Craig Evans**
Assessor:	**Andy Sharpe**
Internal verifier:	**Dan Mills**
Date set:	**12/11/07**
Due date for completion:	**15/02/08**

Unit 1, Assignment 1

Task 1: Engineering sectors and functions

Woodford Marine Diesels Ltd

Woodford Marine Diesels manufactures diesel engines for use on yachts and small boats. The company has been in operation for 15 years and it was started by Steve Woodford and his wife Julia. The company is active in the marine and mechanical secondary (manufacturing) engineering sectors where it designs and manufactures a range of diesel engines.

Maybury Engineering plc

Maybury Engineering is a group of engineering companies based in Midlands and South Wales. The company is organised into three large divisions; aerospace, agricultural, and military vehicles. The company is active in three different secondary engineering sectors: aerospace, mechanical and motor vehicle engineering.

In the aerospace division the company supplies modifications to aircraft for crop spraying and police/coastguard surveillance applications. Aerospace is considered to be a secondary sector, however the crop spraying equipment is used within the agricultural industry which is part of the primary sector and the police/coastguard surveillance equipment could be considered to be part of Telecommunications which is part of the tertiary sector.

In the agricultural division the company manufactures a range of tractors, trailers and agricultural machines such as loaders and bailers. Whilst the manufacture of this equipment clearly lies in the secondary sector the use of the equipment is within the agricultural industry which is part of the primary sector.

In the military division the company manufactures (secondary sector) armoured trucks and personnel transports (based on a standard lorry chassis).

Jet Air Transport Services (JATS) Ltd

Jet Air Transport Services provides aircraft maintenance for large passenger jets (including Boeing 737, 757, 767 and Airbus A300-series aircraft). The company is based at Middleton and Newcastle Airports

JATS provides tertiary engineering maintenance services for aircraft owned and leased by several small regional airlines as well as tour operators (such as JetFly Holidays). At any one time there can be up to seven aircraft undergoing overhaul and maintenance at Middleton and up to three aircraft at Newcastle. The company also supplies ramp engineering services at Newcastle and Leeds airports.

ASSESSOR FEEDBACK FORM

Assessor's comment: Craig – You have correctly identified the engineering sectors of all three companies and in the case of Maybury Engineering this was quite complex. However, you have in most cases stated the activity carried out but have not followed this through into describing the function that they carry out in that sector. Unfortunately at this time you cannot be awarded P1 until you add these descriptions to your answers. For example, you need to describe the design work carried out by Woodford Marine Diesels, or the manufacturing carried out in relation to a particular design and supply of a diesel engine.

Task 2: Company organisation

Woodford Marine Diesels Ltd

The company organisation is a fairly flat structure with five main departmental functions in the company:

- Production (the largest department)
- Development (a small team that develops new products)
- Sales (including two sales engineers)
- Finance (dealing with purchasing and supplies)
- Human resources (dealing with recruitment and training).

All department heads report directly to the Managing Director (Steve Woodford). Company administration is dealt with by Julia Woodford. Personnel issues are handled by the Human Resources manager and a part-time secretary. The company is privately owned and had an annual turnover of £1.8 million in 2007. Woodford Marine Diesels employs about 30 staff and about half of them are engineers and would be considered to be a small to medium enterprise (SME).

The organisational structure (this is a fairly flat but hierarchical structure) is shown in the diagram below:

Woodford Marine Diesels

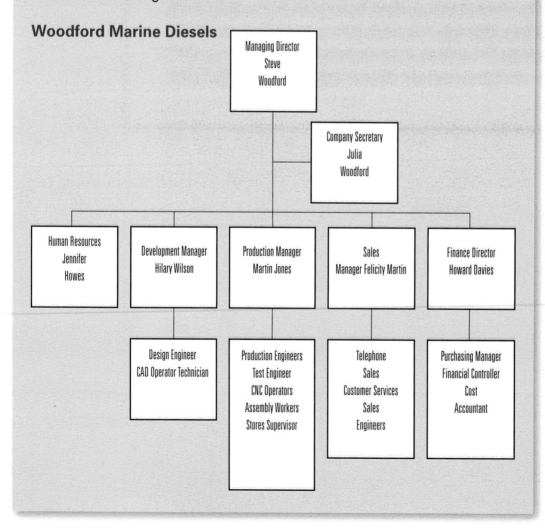

Maybury Engineering plc

Maybury Engineering has a board of directors with a director for each division as well as two non-executive directors and a group chairman who acts as Chief Executive. The group is a large public company and consists of:

- Aerospace Division (located in Coventry). In 2006 this division had an annual turnover of £3.5 million and it currently employs 30 staff
- Agricultural Division (located in Worcester and Evesham). This division had an annual turnover of £11.4 million in 2006 and it employs 125 staff.
- Military Vehicles Division (based in Cardiff). This division had an annual turnover of £18.5 million in 2006 and employs 180 staff.

Each division has its own manufacturing, sales and design team. In addition, some functions are performed at group level (including finance and personnel). The group's head office is located in London. The organisational structure (this is a classic divisional layout which follows a hierarchical structure) is shown below:

Maybury Engineering

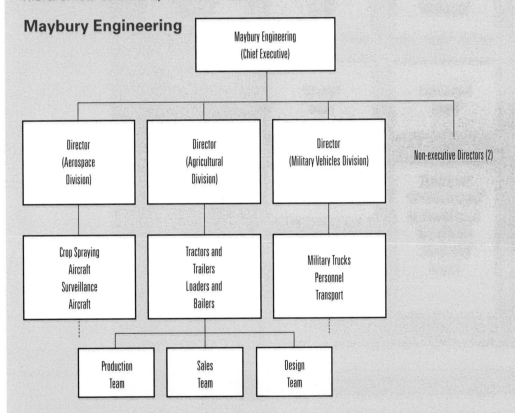

Jet Air Transport Services (JATS) Ltd

All three Team Leaders report to the Engineering manager who then reports to the Chairman. Each Team Leader has several Shift Leaders and each Shift Leader has a number of Engineers, Fitters and other staff reporting to him/her. The company is based at Middleton and Newcastle Airports and it employs 120 full-time and 15 part-time staff. The company is privately owned and the annual turnover was £9.5 million in 2006, most of which was from contracts with JetFly and other operators. JATS has a tall hierarchical structure, as shown below:

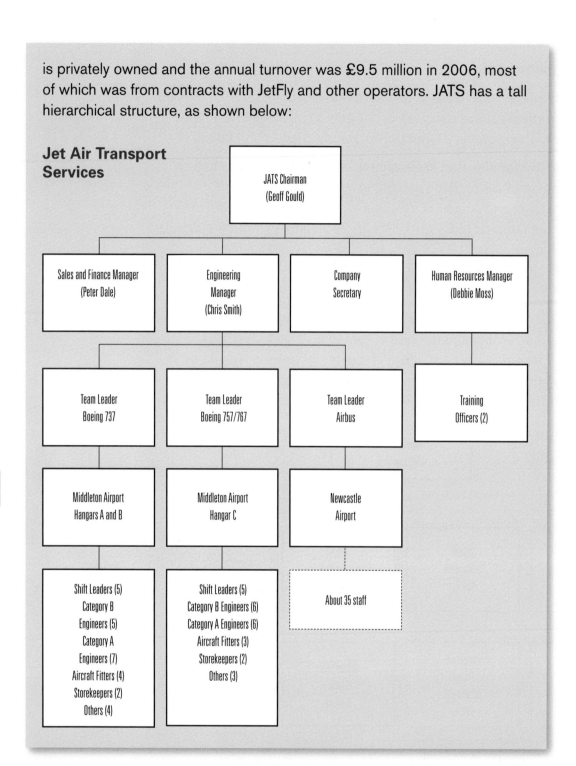

Jet Air Transport Services

JATS Chairman
(Geoff Gould)

Sales and Finance Manager (Peter Dale)

Engineering Manager (Chris Smith)

Company Secretary

Human Resources Manager (Debbie Moss)

Team Leader Boeing 737

Team Leader Boeing 757/767

Team Leader Airbus

Training Officers (2)

Middleton Airport Hangars A and B

Middleton Airport Hangar C

Newcastle Airport

Shift Leaders (5)
Category B
Engineers (5)
Category A
Engineers (7)
Aircraft Fitters (4)
Storekeepers (2)
Others (4)

Shift Leaders (5)
Category B Engineers (6)
Category A Engineers (6)
Aircraft Fitters (3)
Storekeepers (2)
Others (3)

About 35 staff

ASSESSOR FEEDBACK FORM

Assessor's comment: A good answer with some excellent organisational charts. However, your chart for Maybury Engineering could have shown a little more detail and the task would have been a little easier if you had chosen to focus the task on a particular division within the company. It was a good idea to use your work placement at Woodford Marine Diesels for this activity – it's clear that you have learned a lot from this! In each case you have given the size, status and shown the structure. There is ample evidence here to support P2. Well done Craig.

Task 3: Information flow

I have selected JATS to show the information flow.

At JATS a great deal of information is needed in order to perform everyday aircraft maintenance tasks. Engineers have to use a number of different information sources and recording systems and they also need to comply with CAA and company requirements.

Key information sources are:

- Aircraft log (kept in the aircraft)
- Aircraft maintenance manual (supplied by the aircraft manufacturer on CDROM)
- CAA JAR requirements
- Company technical library
- Company quality manual
- Aircraft electronic reporting systems (ECAM and BITE).

The information flow is shown in the diagram below:

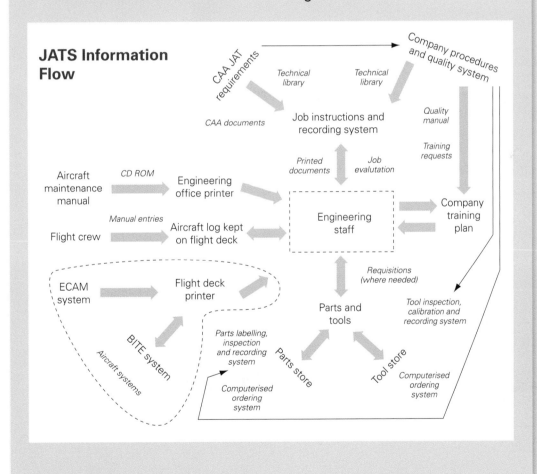

ASSESSOR FEEDBACK FORM

Assessor's comment: Your chart is very confusing – it shows a great deal of detail and the links are not very clear. More evidence is needed to satisfy P3 but, to make things a little easier for you, it would be worth concentrating your efforts on just one type of information and showing how this flows. For example, how a fault message (see printout) generated by the aircraft's ECAM system is dealt with by engineering staff. In relation to this task, you need to show how this particular information is used in order to investigate and eventually rectify the fault. You could use a simple flowchart for this but remember to identify any internal and external systems and people involved. Try to put yourself in the position of an engineer reading the report – how he or she would locate information on Engine 1 Fire Loop A and how this would tell him or her how the Fire Loop should accessed and tested and what equipment is required in order to perform the task. To get started it might be worth taking a look at the relevant sections of the Boeing 757 maintenance manual (the CDROM is in the aircraft workshop technical library). Remember that this is a controlled document – just the same as it would be at JATS! This would enable you to satisfy fully the requirements of P3 in the areas of internal systems, people involved, types of information and work ethics of communications.

MERIT LEVEL ANSWER (ATTEMPTED)

Task 4: Improvements to information flow

JATS could improve information flow in three main ways:

1. Connecting computers together and sharing data between users.
2. Engineers on all three sites could log on to the same computer system.
3. A large electronic screen could be used to replace the workflow boards.

ASSESSOR FEEDBACK FORM

Assessor's comment: There is insufficient evidence provided here for M1. Your answer makes a couple of good suggestions but you've not explained them in terms of the enhancement to functional activities. You need to think about how enhancements could improve the functions within the company. Just "connecting computers together" isn't sufficient as an answer and, although it might be desirable, you've not actually said what the end result and benefits are. I suggest that you do a complete rethink on this and get back to me before you make another start (NB: Your points 1 and 2 are really the same thing!). Finally, point 3 is a nice idea but it would be rather expensive and might not be as visible, easy to use or as reliable as the current manually operated system.

DISTINCTION LEVEL ANSWER (ATTEMPTED)

Task 5: Evaluation of information flow

In order to meet quality standards and CAA-JAR requirements, all work carried out on an aircraft (including the materials and components used) must be fully documented. Because of this, the information flow at JATS is very complex and needs a great deal of accurate recording at each stage in the maintenance cycle.

This chart summarises information flow at JATS. I have highlighted the key information sources that are 'hubs' for information flow.

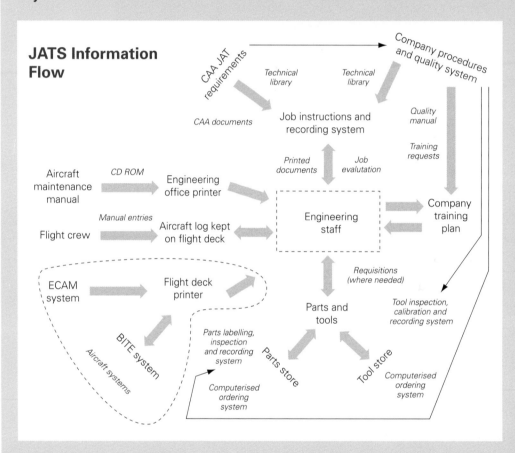

Security

All of the engineering staff are involved in the recording process on a daily basis. They log in at the start of a shift and log out at the end. The system is secure because each engineer has a user name and a password. The company has a written policy on data security and all new members of staff are given a copy of this as part of the company induction.

Verbal and written communication

Information is often passed from one shift to another. Sometimes this happens verbally but more often it requires written messages and notes/ changes made to the workflow boards in the hangar (these are placed so that everybody can see what state the aircraft is in). Special tags and flags are placed on circuit breakers and ground locks to prevent operation when engineers are working on systems that could become unsafe when they are operated or become live. These are hard to miss.

That said, an improvement to this system would be to stipulate that all verbal communication is replaced with or in parallel with written communication, ensuring a full record.

Computer network for central information storage

JATS engineers use stand-alone computers at many stages in their working processes. There are many disadvantages for smooth information flow in such a system and a major improvement would be for JATS to install a central network covering all sites and linking all computers and centralising core information. This would offer the following advantages:

- Improved safety (core data is only required for input once and safety-critical information can be made immediately available to all users as and when required)

- Increased efficiency (engineers on all three sites can log onto the same network, accessing the same information)

- Centralised versions of company procedures could also be accessed easily by all engineering staff

- Aircraft maintenance manuals – provided as CD ROM – could be stored centrally and easily updated.

- CAA JAR requirements can also be amended centrally and the company can be certain that all engineers have access to up-to-date, accurate information.

Workflow boards

The workflow boards are a source of potential errors and inefficiency. Replacing these with electronic equivalents would be a positive move.

Health and safety

Training is given so that all engineers and hangar staff are fully aware of Health and Safety issues. This is important because many hazards are present in the hangars. Signs are placed in prominent places and everybody has to respect them, e.g. when a fluid spill has occurred.

ASSESSOR FEEDBACK FORM

Assessor's comment: This is a good answer, probably just a little shy of achieving D1 in full. It is perfectly acceptable to start with your response to task 3 and to work from there. The diagram shows that you have thought about each stage in turn, what it does and how it could be improved. You have evaluated the information flow at key points and have suggested a number of ways in which improvements could usefully be made. There is one piece of extension from this: you might also want to think about whether each stage of information flow complies with CAA and company requirements. For example, must a record be kept? If so, who needs to keep it, and who should have access to it? Adding that would certainly mean that you had achieved D1.

Unit 6 – Mechanical Principles and Applications
Sample assignment: Static engineering systems

The following is a sample assessment activity. In this case the assignment covers the whole of Learning Outcome 1, but this might not always be the case. You may be assessed separately on the different topics within a Learning Outcome but either way will be given opportunities to meet all of the Pass, Merit and Distinction criteria for the unit.

The grading criteria accessible from this assignment are P1, P2, P3 and M1.

Pass Criteria	Merit Criteria	Distinction Criteria
P1 calculate the magnitude, direction and position of the line of action of the resultant and equilibrant of a non-concurrent coplanar force system containing a minimum of four forces acting in different directions P2 calculate the support reactions of a simply supported beam carrying at least two concentrated loads and a uniformly distributed load P3 calculate the induced direct stress, strain and dimensional change in a component subjected to direct uniaxial loading and the shear stress and strain in a component subjected to shear loading	M1 calculate the factor of safety in operation for a component subjected to combined direct and shear loading against given failure criteria	

Scenario

Whilst working as a junior member of a structural engineering design team, you are asked to investigate the effect of the forces acting on a range of components. You will need to present your findings in a logical and tidy manner that can be easily followed and checked by your team leader.

Task 1

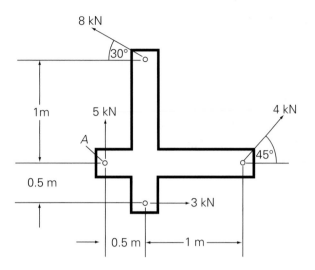

Determine the magnitude, direction and sense of the resultant and equilibrant forces acting on the above component and the perpendicular distance of their line of action from the point A.

(Pass – P1)

Task 2

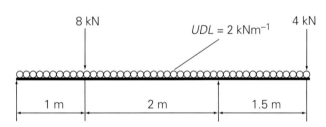

Determine the support reactions for the above simply supported beam.

(Pass – P2)

Task 3

i) A tie bar of length 2.5 m and diameter 10 mm carries an axial load of 12 kN. The modulus of elasticity of the bar material is 180 GPa. Determine the induced tensile stress, the tensile strain and the change in length that occurs.

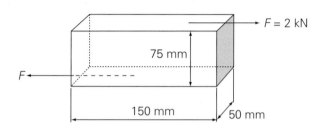

ii) The block of material shown above has a shear modulus of 50 GPa. Determine the shear stress and the shear strain induced by the 2 kN shearing force.

(Pass – P3)

Task 4

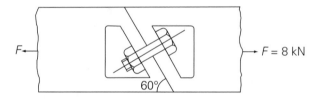

The material for the bolt shown in the angled joint shown above has an ultimate tensile strength of 500 MPa and a shear strength of 300 MPa. The diameter of the bolt is 8 mm. Determine the factor of safety in operation.

(Merit – M1)

PASS LEVEL ANSWERS

Solution to Assignment 1: Static engineering systems

Task 1

This task provides the evidence for you to achieve grading criteria P1 which is:- Calculate the magnitude, direction and position of the line of action of the resultant and equilibrant of a non-concurrent coplanar force system containing a minimum of four forces acting in different directions

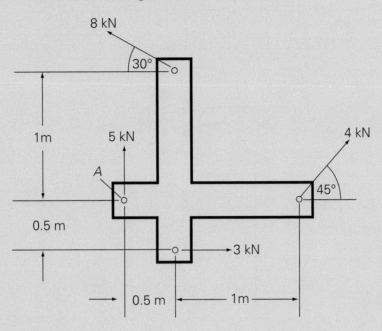

Perhaps the best way to begin this analysis is to draw up a table containing the forces resolved into their horizontal and vertical components, and the moments of the components taken about the point *A*. You should use the following sign conventions to indicate the directions of the force components and moments.

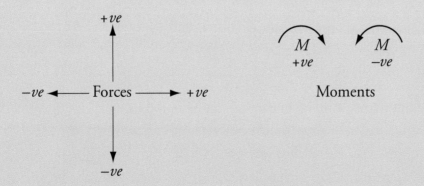

i.e. Horizontal components to the right are positive and to the left are negative Upward vertical components are positive and downward ones are negative Clockwise moments are positive and anticlockwise moments are negative

Force	Horz. Comp. (F_H)	Vert. Comp. (F_V)	Moment of F_H about A	Moment of F_V about A
8 kN	−8 Cos 30 = −6.93 kN	8 Sin 30 = +4.0 kN	−6.93 × 1 = −6.93 kNm	−4 × 0.5 = −.2.0 kNm
4 kN	+4 Cos 45 = +2.83 kN	+4 Sin 45 = +2.83 kN	0	−2.83 × 1.5 = −4.24 kNm
3 kN	+3.0 kN	0	−3 × 0.5 = −1.5 kNm	0
5 kN	0	+5.0 kN	0	0
Totals	Σ_H = −1.1 kN	ΣF_V = +11.83 kN	ΣM_A = −14.67 kNm	

You can now draw a vector diagram showing the sum of the horizontal components and the sum of the vertical components on a force vectors. The sign convention tells you that ΣF_H is to the left and ΣF_v is upward as follows.

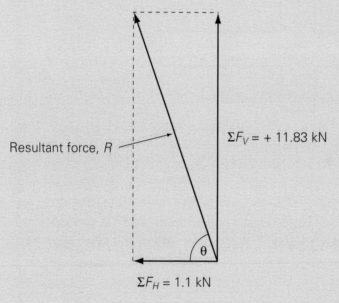

Now use Pythagoras' theorem to find the magnitude of the resultant

$R = \sqrt{(11.83^2 + 1.1^2)}$
$R = 11.88$ kN

The angle that the resultant makes with the horizontal can be found using simple trigonometry.

$Tan\ \theta = \dfrac{11.83}{1.1} = 10.75$

$\theta = 84.7°$

Finally you need to find the perpendicular distance a, of line of action of resultant and equilibrant from point A.

136

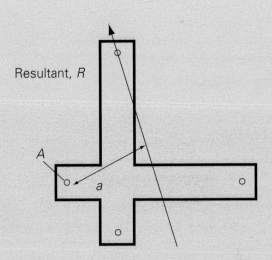

Resultant, R

A

a

Because the sum of the moments is negative i.e. anticlockwise about A, the line of action of the resultant must be to the right of A as shown above. The distance a

Moment of resultant about A = Sum of moments about A

$$11.88 \times a = 14.67$$
$$a = 1.23 \text{ m}$$

The equilibrant of the system will be equal in magnitude to the resultant but acting in the opposite direction.

ASSESSOR FEEDBACK FORM

Commentary on the work

Note that the forces have been correctly resolved and Pythagoras' theorem has been applied to find the resultant and equilibrant.

The findings are presented in a logical manner that is easy to follow and check.

The values for the resultant and its direction are correct. The reasoning about the position of the line of action of the resultant is valid and you have obtained the correct distance of the line from the point A.

This means that P1 can be awarded.

Task 2

This task provides the evidence for you to achieve the grading criteria P2 which is:- Calculate the support reactions of a simply supported beam carrying at least two concentrated loads and a uniformly distributed load

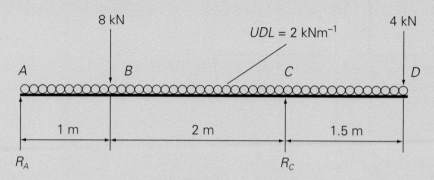

One of the conditions for static equilibrium is that that the sum moments of the forces acting on a body must be zero. Sometimes this is also called the principle of moments and you always use it to find the support reactions of *a* simply supported beam

Begin by taking moments about A to find R_c

$$\Sigma M_A = 0$$
$$(8 \times 1) + (4 \times 4.5) + (2 \times 4.5 \times 2.25) - (R_C \times 3) = 0$$
$$8 + 18 + 20.25 - 3R_c = 0$$
$$46.25 = 3R_c$$
$$R_C = 46.25/3 = 15.42 \text{ kN}$$

Now take moments about C to find R_A

$$(R_A \times 3) + (4 \times 1.5) + (2 \times 1.5 \times 0.75) - (8 \times 2) - (2 \times 3 \times 1.5) = 0$$
$$3R_A + 6 + 2.25 - 16 - 9 = 0$$
$$3R_A = 16.75$$
$$R_A = 16.75/3 = 5.52 \text{ kN}$$

Finally check that upward forces are equal to downward forces, which is also a condition of static equilibrium.

Sum of upward forces = 15.423 + 5.52 = 21 kN
Sum of downward forces = 8 + 4 + (2 × 4.5) = 21 kN

ASSESSOR FEEDBACK FORM

Commentary on the work

This would get us to a correct value for R_c. A common error could occur when taking moments about C. Remember that the load is uniformly distributed. It acts on both sides of C, and each part will have its own moment, which has to be included in the working.

It is useful to check that the sum of the downward load equals the sum of the two support reactions.

This means that P2 can be awarded.

Task 3

This task provides the evidence for you to achieve the grading criteria P2 which is:- Calculate the induced direct stress, strain and dimensional change in a component subjected to direct uniaxial loading and the shear stress and strain in a component subjected to shear loading

i)   The first part of the task is a calculation for a loaded tie bar, i.e. a component loaded in tension.

Begin by calculating the cross-sectional area of tie bar

$$A = \frac{\pi d^2}{4} = \frac{\pi \times 0.01^2}{4}$$
$$A = 78.5 \times 10^{-6} \text{ m}^2$$

You can now find the tensile stress in tie bar

$$\sigma = \frac{F}{A} = \frac{12 \times 10^3}{78.5 \times 10^{-6}}$$

$$\sigma = 153 \times 10^6 \text{ Pa or } 153 \text{ MPa}$$

The next thing is to find the tensile strain in tie bar

$$E = \frac{\sigma}{\varepsilon}$$

$$\varepsilon = \frac{\sigma}{E} = \frac{153 \times 10^6}{180 \times 10^9}$$

$$\varepsilon = 850 \times 10^{-6}$$

Finally you can find the change in length of bar

$$\varepsilon = \frac{x}{l}$$

$x = l\varepsilon = 2.5 \times 850 \times 10^{-6}$

$x = 2.13 \times 10^{-3}$ m or 2.13 mm

ii) The second part of this task is a calculation for a block of material loaded in shear.

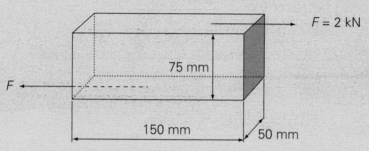

Begin by calculating the sheared cross-sectional area of block

$A = 150 \times 10^{-3} \times 50 \times 10^{-3}$
$A = 7.5 \times 10^{-3}$ m²

You can now find the shear stress in block

$$\tau = \frac{F}{A} = \frac{2 \times 10^3}{7.5 \times 10^{-3}}$$

$\tau = 267 \times 10^3$ Pa or 267 kPa

Finally you can find the shear strain in block

$$G = \frac{\tau}{\gamma}$$

$$\gamma = \frac{\tau}{G} = \frac{267 \times 10^3}{50 \times 10^9}$$

$\gamma = 5.34 \times 10^{-6}$

MERIT LEVEL ANSWER

Task 4

This task provides evidence for you to achieve the grading criteria M1:-
Calculate the factor of safety in operation for a component subjected to
combined direct and shear loading against given failure criteria

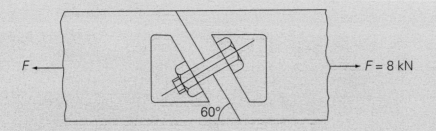

Begin by finding the cross-sectional area of the bolt

$$A = \frac{\pi d^2}{4} = \frac{\pi \times 0.008^2}{4}$$
$$A = 50.3 \times 10^{-6} \text{ m}^2$$

Now find the tensile force F_T acting on the bolt by resolving the force F in a
direction perpendicular to the joint face.

$F_T = \text{F Sin } 60 = 8 \times 10^3 \times 0.866$
$F_T = 6.93 \times 10^3 \text{ N}$

You can now find the tensile stress in the bolt.

$$\sigma = \frac{F_T}{A} = \frac{6.93 \times 10^3}{50.3 \times 10^{-6}}$$

$\sigma = 138 \times 10^6 \text{ Pa or } 153 \text{ MPa}$

The next thing is to find the factor of safety against tensile failure.

$$FOS = \frac{\text{Ultimate tensile stress}}{\text{Actual tensile stress}} = \frac{\text{UTS}}{\sigma}$$

$$FOS = \frac{500 \times 10^6}{138 \times 10^6}$$

$FOS = 3.62$

You can now turn your attention to the shear force F_S acting on the bolt. This is
found by resolving the force F in a direction parallel to the joint face.

$F_S = \text{F Cos } 60 = 8 \times 10^3 \times 0.5$
$F_T = 4.0 \times 10^3 \text{ N}$

You can now find the tensile stress in the bolt.

$$\tau = \frac{F_s}{A} = \frac{4.0 \times 10^3}{50.3 \times 10^{-6}}$$

$\tau = 79.5 \times 10^6$ Pa or 79.5 MPa

Finally, you can find the factor of safety against shear failure

$$FOS = \frac{\text{Shear strength}}{\text{Actual shear stress}} = \frac{\text{Shear strength}}{\tau}$$

$$FOS = \frac{300 \times 10^6}{79.5 \times 10^6}$$

$FOS = 3.77$

The smaller of the two factors of safety values, i.e. the factor in tension, is the one in operation for the joint.

i.e. Factor of safety in operation = 3.62

ASSESSOR FEEDBACK FORM

Commentary on the work

At this stage, common errors could appear when resolving the force perpendicular and parallel to the joint face. Ensure that you use the cosine and sine rules in the right places, and always stop and think if you are selecting the correct operation.

This means that M1 can be awarded.